NEMESIS

NEMESIS

THE DEATH STAR

The Story of a Scientific Revolution

Richard Muller

Introduction by Dr Luis Alvarez

HEINEMANN : LONDON

William Heinemann Ltd
Michelin House, 81 Fulham Road, London SW3 6RB

LONDON   MELBOURNE   AUCKLAND

First published in Great Britain in 1989
Copyright © 1988 by Richard Muller

The author gratefully acknowledges permission to reprint the map on page 118,
which was originally published by The Geological Society of America on page 29
of Special Paper 190, titled "Geological Implications of Impacts of Large Asteroids
and Comets on Earth," edited by L. T. Silver and P. H. Schultz. Map © Richard
A. F. Grieve, reprinted by permission of Richard A. F. Grieve.

British Library Cataloguing in Publication Data

Muller, Richard
Nemesis.
1. Organisms. Extinction
I. Title
575'.7

ISBN 0-434-48161-0

Printed in Great Britain by
Redwood Burn Limited, Trowbridge, Wiltshire

To Luie
for teaching me the art of physics
and much much more

Acknowledgments

This book is a direct descendant of my article, "An Adventure in Science," which was written for the *New York Times Magazine* at the request of Harvey Shapiro. I am particularly grateful to Luis Alvarez, Walter Alvarez, and Frank Asaro for their care in reviewing early versions of the manuscript and for their help in filling the gaps in my knowledge. Very helpful suggestions were received from Bill Press, Piet Hut, Ronald Kahn, Terry Mast, Gordon Getty, and Dudley Wright. I thank Rosemary, Betsy, and Melinda for their patience as I hacked away at the word processor, and Apple Computer for making hacking on a MacIntosh so much fun.

Contents

Part 2. Nemesis

Introduction

THIS BOOK is not a scientific autobiography, as it may appear to someone casually thumbing through it. It is a description of research on one of the most exciting and revolutionary problems in modern science, but it is more than that. I have long felt that most books about science, indeed even most scientific autobiographies, neglect to mention what I would most like to know—how scientists happened to be working on their most important problems and what the nature of their thought processes was as they attacked those problems. I am convinced that every student of physics will read and reread *Nemesis* several times, learning important lessons on each occasion as well as having a wonderful time. And I think the lay reader who has always wondered about the same question, how is science really done, will find the book equally interesting. This book lets the nontechnical person get inside the mind of the scientist and see what the experience of discovery and invention is really like. I know of no other book about physics that succeeds as well as *Nemesis* in describing the art of science. I believe that the lay reader will have difficulty in putting the book down and that as several of my nontechnical friends have done, they will stay up all night to finish it.

Rich Muller is certainly the best student I have had in my fifty-five years of physics research, but I also believe that he is the best wide-ranging experimental physicist in his generation. In fact, he may be the equal of the best broad experimentalists that I have known in my career. Most physicists today achieve their distinction in one narrow field, but Rich has managed to make major contributions in many. He has that rare talent that enables him to improvise and think up new ways of doing science, and the leadership to put together teams to follow his ideas. Rich's group discovered that the microwaves from the Big Bang made a "great cosine in the sky" and showed that the earth and the Milky Way galaxy move through space at about one million miles per hour. His "Image Sharpness Theorem" for optics explains how the eye can tell when it is in focus, and gives a method to correct for atmospheric distortion in telescopes. Rich invented a method of dating that is now widely used in archeology and geology, and is thousands of times more sensitive than the previous method. Recently he led a team that developed an automated system that discovered several supernovas. From this list, you would not guess that Rich earned his Ph.D. working in elementary particle physics. He has spent every summer for the last fifteen years working on national security problems, and for the past four years has been making important contributions in arms control by participating on a committee of the National Academy of Sciences that meets with the Soviet Academy twice per year, in Moscow and in Washington, D.C. But his most important work so far may be the work he describes in this book.

Rich's book gives the best description that has been written of the development of the Nemesis story, from the original discovery of the extraterrestrial impact by my son Walt, Frank Asaro, Helen Michel, and myself, to the ongoing search for the star by Rich and his colleagues. The book reads like a mystery novel, because the scientific process is largely the unraveling of mysteries and the work of a scientist is largely that of a detective. But the process of answering "whodunit?" is much harder when many people believe that the perpetrator was long ago sent to prison or perhaps that no crime has been committed. Rich's book also makes a vivid analogy with the work of an explorer, "trying to put together a map of an unknown world, unsure of the value of what he is going to find and how he is going to repay his debts, while suffering from shortages of supplies and attacks of the natives."

I am now fully convinced that the periodicity that paleontologists discovered in the mass extinctions is real and that Rich's Nemesis theory is the only plausible explanation, although we won't have proof until the star is found. The discovery of Nemesis would force a complete rewrite of our picture of the solar system, of our picture of evolution, and even more.

I recognize that the newspapers, perhaps even most scientists, regard this area of science as extremely controversial. I think that is because they are not familiar with all the compelling facts and because the revolutionary nature of the discoveries goes against much of the dogma (but not against the observations) of present-day science. The early establishment of consensus is not to be expected in a scientific revolution. When Rich finds Nemesis (and I believe that he will), the revolution will complete itself very rapidly, and those who have read this book will be able to share in the full thrill, the thrill that is the ultimate goal of much scientific work, the thrill of turning past dogma upside down and opening a new world with unknown frontiers.

Luis W. Alvarez
December 22, 1987

Part 1

THE CRETACEOUS CATASTROPHE

1. Cosmic Terrorist

LUIS ALVAREZ walked into my office looking like he was ready for a fight. "Rich, I just got a crazy paper from Raup and Sepkoski. They say that great catastrophes occur on the Earth *every* 26 million years, like clockwork. It's ridiculous."

I recognized the names of the two respected paleontologists. Their claim *did* sound absurd. It was either that, or revolutionary, and one recent revolution had been enough. Four years earlier, in 1979, Alvarez had discovered what had killed the dinosaurs. Working with his son Walter, a geologist, and Frank Asaro and Helen Michel, two nuclear chemists, he had shown that the extinction had been triggered 65 million years ago by an asteroid crashing into the Earth. Many paleontologists had initially paid no attention to this work, and one had publicly dismissed Alvarez as a "nut," regardless of his Nobel Prize in physics. Now, it seemed, the nuts were sending their theories to Alvarez.

"I've written them a letter pointing out their mistakes," Alvarez continued. "Would you look it over before I mail it?"

It sounded like a modest request, but I knew better. Alvarez expected a lot. He wanted me to study the "crazy paper," understand it in detail, and then do the same with his letter. He wanted each of his calculations redone from scratch. It would be a time-consuming and tedious task, but I couldn't turn him down. He and I depended on each other for this kind of

3

work. We knew we could trust each other to do a thorough job. Moreover, we had enough mutual respect so that we didn't mind looking foolish to each other, although neither of us liked looking foolish to the outside world. So I reluctantly accepted the task, as I had many times before.

The Alvarez theory had slowly been gaining acceptance in the scientific world. The astronomers had been the most receptive, perhaps because their photographs often showed large asteroids and comets floating around in space in orbits that crossed the path of the Earth. They knew that disastrous impacts must have taken place frequently in the past. Many geologists had likewise been won over. But a majority of paleontologists still seemed opposed to the theory, which was disruptive to their standard models of evolution. Alvarez took pride in the fact that some of the most respected paleontologists nevertheless liked his theory, including Stephen Jay Gould, Dale Russell, David Raup, and J. John Sepkoski.

I began my assignment by reading the paper by Raup and Sepkoski. They had collected a vast amount of data on family extinctions in the oceans, far more than had previously been assembled. That fact disturbed me; I hate to dismiss the conclusions of experts, especially conclusions based on such minute study. Their analysis showed that there were intense periods of extinctions every 26 million years. It wasn't surprising that there should be extinctions this often, but it *was* surprising that they should be so regularly spaced. Alvarez's work had shown that at least two of these extinctions were caused by asteroid impacts, the one that killed the dinosaurs at the end of the Cretaceous period, 65 million years ago, and the one that killed many land mammals at the end of the Eocene, 35–39 million years ago. (The age was uncertain because of the difficulty of dating old rocks.)

Astrophysics was a field I thought I knew; my work in it had earned me a professorship in physics at Berkeley and three prestigious national awards. But the paper beggared my understanding. I found it incredible that an asteroid would hit precisely every 26 million years. In the vastness of space, even the Earth is a very small target. An asteroid passing close to the sun has only slightly better than one chance in a billion of hitting our planet. The impacts that do occur should be randomly spaced, not evenly strung out in time. What could make them hit on a regular schedule? Perhaps some cosmic terrorist was taking aim with an asteroid gun. Ludicrous results require ludicrous theories.

I hurried to the end of their paper, like a reader cheating on a mystery novel, to see how Raup and Sepkoski would explain the periodicity. I was disappointed to find that they had no theory, only facts. Physicists have a wry saying: "If it happens, then it must be possible." Many discoveries had been missed because scientists ignored data that didn't fit into their established mode of thinking, their paradigm, and I didn't want to fall into that trap. Maybe it would be best to review their data, I thought, and try to judge them independently of theory. On a chart, they had plotted the varying extinction rate for the last 250 million years. The big peaks in the rate were spaced 26 million years apart.

Next I turned to Alvarez's letter. He thought there were several mistakes in the way that Raup and Sepkoski had analyzed their data. Several of the apparent peaks, he argued, should be removed from the analysis because of their low statistical certainty. Likewise, both the Cretaceous and Eocene extinctions should not be considered as part of a periodic pattern, since they were due to asteroid impacts and therefore must be random in time. This had been as obvious to Alvarez as to me. With these extinctions removed, the remaining ones were so widely separated that it looked like all evidence for periodicity had vanished.

Alvarez's approach was convincing, but was it right? It was my job to be the devil's advocate, to defend the conclusions of Raup and Sepkoski. I went back to their paper and looked at the chart again. I mustn't be too skeptical, I thought. I replotted their data, substituting the conventions of physicists for those of paleontologists. I gave each extinction an uncertainty in age as well as in intensity. The new chart looked more impressive than I had expected. It was a rough version of the one shown on page 6.

I had placed the arrows at the regular 26-million-year intervals. Eight of them pointed right at extinction peaks; only two missed. The peaks certainly seemed to be evenly spaced.

Maybe they were right. I realized I had better reexamine Alvarez's case, and see if *it* was flawed. This job was turning out to be more fun than I had expected.

On my second reading of Alvarez's letter, I found it particularly dubious that the Cretaceous and Eocene extinctions should be excluded. How do we know that asteroids do not hit the Earth periodically? I asked. Maybe our failure to arrive at a theory just meant that we hadn't been clever enough. Not finding something is not the same as proving it is not there. I decided to reserve judgment.

A few minutes later Alvarez stopped by to see if I had finished, and I told him that I had found a mistake in his logic. It had been improper to exclude the Cretaceous and Eocene mass extinctions, I said. I presented my case like a lawyer, interested in proving my client innocent, even though I wasn't totally convinced myself.

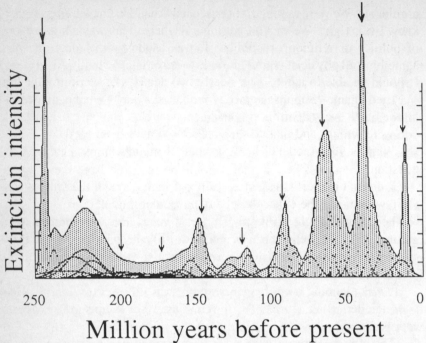

XBL 8611-4774

The data of paleontologists Raup and Sepkoski that show great mass extinctions occurring every 26 million years, as replotted by the author. This plot inspired the Nemesis theory.

Alvarez rejoined strongly, like a lawyer himself. "To keep those extinctions in the analysis would be cheating," he said. His belligerent offense threw me momentarily off balance. "You're taking a no-think approach," he continued. "A scientist is not allowed to ignore something he knows to be true, and we know those events were due to asteroid impacts."

I knew Alvarez far too well to acquiesce in his onslaught. My approach was not no-think, I said. It was proper to ignore certain "prior knowledge" in testing a hypothesis. He had no right to assume that the Cretaceous and Eocene extinctions could not be a part of a larger periodic pattern. Maybe if we were clever enough to find the right explanation, we would see that asteroid impacts can be periodic.

Alvarez repeated his previous argument, with a little more emphasis on the phrase "no-think." His body language seemed to say, "Why doesn't Rich understand me? How can he be so dumb?" I repeated my old arguments. We were talking right past each other. He knew he was right. I knew I was right. We weren't getting anywhere. This was not a question of politics or religion or opinion. It was a question of data analysis, something all physicists should be able to agree on. Certainly Alvarez and I should be able to agree, after nearly two decades of working together.

I tried again. "Suppose someday we found a way to make an asteroid hit the Earth every 26 million years. Then wouldn't you have to admit that you were wrong, and that all the data should have been used?"

"What is your model?" he demanded. I thought he was evading my question.

"It doesn't matter! It's the possibility of such a model that makes your logic wrong, not the existence of any particular model."

There was a slight quiver in Alvarez's voice. He, too, seemed to be getting angry. "Look, Rich," he retorted, "I've been in the data-analysis business a long time, and most people consider me quite an expert. You just can't take a no-think approach and ignore something you know."

He was claiming authority! Scientists aren't allowed to do that. Hold your temper, Rich, I said to myself. Don't show him you're getting annoyed.

"The burden of proof is on you," I continued, in an artificially calm voice. "I don't have to come up with a model. Unless you can demonstrate that no such models are possible, your logic is wrong."

"How could asteroids hit the Earth periodically? What is your model?" he demanded again. My frustration brought me close to the breaking point. Why couldn't Alvarez understand what I was saying? He was my scientific hero. How could *he* be so stupid?

Damn it! I thought. If I have to, I'll win this argument on *his* terms. I'll invent a model. Now my adrenaline was flowing. After another moment's thought, I said: "Suppose there is a companion star that orbits the sun. Every 26 million years it comes close to the Earth and does something, I'm not sure what, but it makes asteroids hit the Earth. Maybe it brings the asteroids with it."

I was surprised by Alvarez's thoughtful silence. He seemed to be taking the idea seriously and mentally checking to see if there was anything wrong with it. His anger had disappeared.

Finally he said, "You surprised me, Rich. I was sure you would come up with a model that brought in dust or rocks from *outside* the solar system, and then I was going to hit you with a fact I bet you didn't know, that the iridium layer associated with the disappearance of the dinosaurs came from *within* our own solar system. The rhenium-187/rhenium-185 ratio in the boundary clay is the same as in the Earth's crust. I figured that you didn't know this. But your companion star was presumably born along with the sun, and so it would have the same isotope ratios as the sun. The argument I was holding in reserve is no good. Nice going."

Alvarez paused. He had been trying to think a step ahead of me, anticipating my moves, like a chess master. He had guessed what my criticism would be and had his answer ready—but I had made a different move. He seemed pleased that his former student could surprise him. He finally said, "I think that your orbit would be too big. The companion would be pulled away by the gravity of other nearby stars."

I hadn't expected the argument to cool down so suddenly. We were back to discussing physics, not authority or logic. I hadn't meant my model to be taken that seriously, although I had felt that my point would be made if the model could withstand assault for at least a few minutes. He was right that I was ignorant of the rhenium discovery. Alvarez's son Walter, a geologist, had found a clay layer that had been deposited in the oceans precisely at the time the dinosaurs were destroyed. This clay layer, the elder Alvarez hypothesized, had been created by the impact of an extra-terrestrial body (such as a comet or an asteroid) on the Earth. Rhenium comes in several forms—among others, rhenium-185, which is stable, and rhenium-187, which is radioactive and disappears with a half-life of 40 billion years. In the 4.5 billion years since the formation of the solar system, approximately 8% of the rhenium-187 should have disintegrated. And, in fact, roughly that amount had. Unless the rhenium in the clay had been produced at the same time as the rhenium in the Earth (i.e., at the formation of the solar system), the ratios were very unlikely to be so nearly identical. In other words, the extraterrestrial body would appear to have been born at the same time as the sun.

Now I took the initiative. "Let's see if you are right that the star would be pulled away from the sun. We can calculate how big the orbit would be." I wrote Kepler's laws of gravitational motion on the blackboard. The major diameter of an elliptical orbit is the period of the orbit, in this case 26 million years, raised to the 2/3 power, and multiplied by 2. My Hewlett-Packard 11C pocket calculator quickly yielded the answer: 176,000 astronomical units, i.e., 176,000 times as far as the distance from the Earth to the sun, about 2.8 light-years. (A light-year is the distance that light travels in one year.) That put the companion star close enough to the sun so it would not get pulled away by other stars. Alvarez nodded. The theory had survived five minutes, so far.

"It looks good to me. I won't mail my letter." Alvarez's turnaround was as abrupt as his argument had been fierce. He had switched sides so quickly that I couldn't tell whether I had won the argument or not. It was my turn to say something nice to him, but he spoke first. "Let's call Raup and Sepkoski and tell them that you found a model that explains their data."

So was born the Nemesis hypothesis, though I had no idea at the time where this would lead me.

2. Disaster

EXCEPT FOR the confusing new data supplied by Raup and Sepkoski, our understanding of what happened 65 million years ago had seemed nearly complete. The Alvarez team and teams of other scientists around the world had painted a fairly complete picture of the disaster. There were still some details that weren't fully understood, but the broad picture can be summarized as follows.

At the end of the Cretaceous period, the golden age of dinosaurs, an asteroid or comet about 5 miles in diameter (about the size of Mt. Everest) headed directly toward the Earth with a velocity of about 20 miles per second, more than 10 times faster than our fastest bullets. Many such large objects may have come close to the Earth, but this was the one that finally hit. It hardly noticed the air as it plunged through the atmosphere in a fraction of a second, momentarily leaving a trail of vacuum behind it. It hit the Earth with such force that it and the rock near it were suddenly heated to a temperature of over a million degrees Celsius, several hundred times hotter than the surface of the sun. Asteroid, rock, and water (if it hit in the ocean) were instantly vaporized. The energy released was greater than that of 100 million megatons of TNT, 100 teratons, more than 10,000 times greater than the total U.S. and Soviet nuclear arsenals.

Before a minute had passed, the expanding crater was 60 miles across and 20 miles deep. (It would soon grow even larger.) Hot vaporized material from the impact had already blasted its way out through most of the atmosphere to an altitude of 15 miles. Material that a moment earlier had been glowing plasma was beginning to cool and condense into dust and rock that would be spread worldwide. The entire Earth recoiled from the impact, but only a few hundred feet. The length of the year changed by a few hundredths of a second.

The deep crater may have reached through the crust of the Earth to the mantle. The rock at this depth is very hot due to the natural radioactivity of trace amounts of potassium, uranium, and thorium. The hot rock had turned to liquid as soon as the weight of the rock above it was removed. Great pressure from the Earth's interior quickly filled in most of the crater with melted rock from below. It is possible that molten rock continued to flow in a great outpouring. This is not known for certain, because the impact site has not been definitely identified. One suggested location is at the Deccan traps in India, one of the greatest outpourings of deep basaltic rock that has taken place in the last billion years of the Earth's history. Two hundred thousand square miles of the Earth were covered when the Deccan traps were formed, and this is known to have taken place at about the right time, approximately 65 million years ago.

Shock waves from the impact rattled the Earth with energy much greater than that of the largest earthquakes humans have experienced, probably a million times more energetic than the one that devastated San Francisco in 1906. The shock may have triggered other earthquakes, causing aftershocks lasting for months or years. This is speculation, for we don't know enough about the crust of the Earth to say with any certainty. Weak points in the Earth's crust may have opened and sprouted new volcanoes. It is impossible to guess all the effects, or how long they lasted.

Our failure to find the crater suggests either that it was obscured by the subsequent outflow of magma or that the impact occurred in the two thirds of the Earth that is covered by ocean. According to the latter scenario, the asteroid quickly punched through the water layer much as a rock would through a puddle, since in most places the depth of the ocean is less than the hypothesized 5-mile diameter of the asteroid. The splash created a great tsunami, or tidal wave, which grew hundreds of feet high as it swept toward shore. The giant wave circled the Earth many times, inundating the coastal regions. Only the interiors of the continents were spared.

Rock from the crater, mixed with vaporized comet material, cooled to several thousand degrees Celsius as it rose in a fireball up through the atmosphere. It was then about as hot and bright as the surface of the sun, the greatest fireball ever seen by living creatures. But they didn't watch it for long, because the intense heat radiated from the glowing cloud burned everything within sight. The heat of the fireball also fused nitrogen and oxygen in the air to make nitrous oxides, a constituent of modern smog. Some of this gas later combined with water in the atmosphere to make nitric acid. Likewise, sulfur dioxide from burning plant material formed sulfuric acid, which together with the nitric acid eventually fell to Earth. This acid rain may have been strong enough to dissolve the shells of creatures that lived in the surface waters of the oceans.

Some of the material ejected from the crater flew out of the atmosphere in ballistic trajectories and, like ICBMs (intercontinental ballistic missiles), rained havoc on distant continents. Some escaped into space, perhaps to hit the Earth on an anniversary of the impact when both the Earth and the ejecta returned to the same part of their orbits around the sun. As the fireball reached the top of the atmosphere, it bobbed like a cork, floating on the cooler air beneath it, but it had nothing to hold it together and it began to spread out over the entire globe. As it spread, its color cooled from a glowing red to an impenetrable black. The surviving creatures below probably thought that night had come early, but it was a night without a moon, without stars. The dinosaurs could not see their own claws in front of their faces. Morning would not come for several months.

A few animals had avoided the initial destruction, and at first they seemed to manage surprisingly well. Most of the plant eaters could still find food, although the settling dust added a gritty texture to it all. Some of the carnivores were accustomed to hunting in the dark, although they had never experienced blackness such as this. But the ultimate source of all food, the sun, had been effectively blocked out. Without sunlight, there was no photosynthesis, no creation of sugar and starch from carbon dioxide and water. Unseen by the animals, the plants were turning from green to yellow, and then to brown. If not for the darkness, it would have been a beautiful autumn scene. The larger herbivores began to starve, followed soon afterward by the large carnivores. Similar death occurred in the oceans. Phytoplankton, the first link in the food chain for the oceans, died from the acid rain and lack of sunlight, and the higher organisms quickly followed.

Without sunlight, the temperature of much of the Earth began to drop. On much of the land the temperature soon fell below freezing. Only those fortunate few animals that had already begun to hibernate didn't notice. Warm-blooded creatures had an advantage in their ability to endure the cold, but they also required more food. The coastal regions had temperatures moderated by the oceans, but these were the areas that had been devastated by the tsunami. Tremendous storms were generated by the great temperature differences between the oceans and the land masses. The storms rained the nitrous and sulfuric acids onto land and sea.

The tiny particles of dust high in the atmosphere began to stick to each other, agglomerating into larger particles, which fell to Earth more quickly. All over the Earth the dust settled to form a layer about a half-inch thick. (One day Walter Alvarez would puzzle over a section of this layer in an outcropping near Gubbio, Italy, where it had been brought up to the surface as the crust of the Earth folded during the creation of the Apennine mountains. He would cut a piece of it out to give as a gift to his father.)

As the dust settled, sunlight began to filter through to the Earth's surface. Virtually every animal and plant had died. The plants were probably the first to revive. Many spores and seeds had been left behind. Life is incredibly robust. For some plants, the three-month period of darkness and cold was no worse than a severe winter; these resprouted from seeds and roots. It was a miracle that any of the higher life forms made it through. Indeed, almost every land animal that weighed more than about 50 pounds had been extinguished, probably because they were most vulnerable in their high position on the food chain. The reptilian dinosaurs, both on land and in the sea, had vanished forever. The nearest relatives of the dinosaurs to survive were the birds. We don't really know why. Perhaps it was their mobility, their ability to search for warmth and food. Perhaps they were better able than other dinosaurs to feed off decaying matter and seeds. Perhaps it was their warm blood. We can't be sure because we don't even know if the other dinosaurs were warm-blooded or cold-blooded. There was no obvious pattern that explained why certain creatures had survived, and others had not. Life is very robust, and yet very fragile.

With plant life suddenly sprouting all around, the few creatures that had made it through the catastrophe found themselves in a virtual Garden of Eden. Their natural enemies were gone, and food from plants was abundant. However, their species would not continue unless they could find mates. Those that did quickly spread out over the Earth, like (much more recently) the rabbits let loose in Australia. They filled the ecological niches that had been denied them by their previously "fitter" enemies. The slate of evolution had been wiped nearly clean. Now there was plenty of room for Nature to try new inventions. In fact, the great catastrophe was marked in the paleontological record not only by the disappearance of species, but also by the great proliferation of new species that followed. Like a forest fire clearing the brush, the destruction may have been necessary to clear the world of weak creatures that had nevertheless held on to a niche simply by virtue of filling it before any other animal. In evolution, possession really is nine tenths of the law. Survival of the fittest had been hindered by an equally powerful principle: survival of the first. The catastrophe had cleared out whatever stagnation there had been in the process of evolution. Once again there was room for free experimentation.

In this picture, evolution wasn't exactly what it was once thought to be, creatures fighting other creatures to determine which were the fittest. To survive, creatures also had to be adapted to endure catastrophe. Maybe that's why the mammals made it through. They had put up with trauma for 100 million years, the trauma of attempting to live with the dinosaurs.

When Luis Alvarez's team first analyzed the climatic effects of the impact, it could not decide whether the dust surrounding the Earth would cause the surface temperature to drop or rise. The answer depended on which effects were most important. Blocking the sunlight should cause the Earth to chill, they thought at first. But high dust is better at absorbing sunlight than clouds, which reflect most of the light that hits them. If the dust became hot enough, it could radiate the heat to the surface of the Earth, warming it up. Geologist Eugene Shoemaker had pointed out that if the asteroid hit in the ocean, the injection of water vapor into the atmosphere could increase the "greenhouse effect" and warm the climate. It was hard to know which effect would dominate, whether the temperature would rise or drop. This question had interested Brian Toon, a physicist who worked at NASA's Ames Research Laboratory in Sunnyvale, California. Toon and his colleagues were eventually led to an amazing conclusion, one that had immediate practical consequences far afield from the mystery of the dinosaurs. (A few years later this work was under extensive study by the U.S. Department of Defense.)

Toon and his colleagues recognized that the dust cloud would cause both the absorption and reemission of the solar radiation to take place at higher altitudes than usual. Normally the surface of the Earth receives light radiation from the sun and roughly an equal amount of infrared radiation from the upper atmosphere. With the dust layer stopping the direct sunlight, the land would receive only half as much energy. Normally the temperature of an object bathed in radiation is proportional to the fourth root (i.e., the square root of the square root) of the energy hitting it. Since the surface temperature of the dust-free Earth is normally about 28 degrees Celsius (300 degrees above absolute zero), a drop could be expected, on an absolute scale, of the fourth root of 2, a factor of 1.18, down to 252 degrees above absolute zero, about 20 degrees below freezing. More detailed computer simulations that took into account the fact that the atmosphere has several effective layers essentially verified the simple calculation, although the temperature drop was not quite as severe.

If snow covers a large fraction of the Earth, then even the return of sunlight may not warm up the Earth. Although it can begin to reach the ground once the dust settles, the sunlight will be reflected by the white snow before it can turn into infrared heat. It may take something external, such as another impact or a volcanic eruption that would spread dust over the snow, to make the Earth absorb sunlight again. Some paleontologists had believed that it was climate change that had killed the dinosaurs. But it now seemed likely that one was not a consequence of the other, but that in fact both had been brought about by the same cause, the asteroid impact. Perhaps a climate change *had* been the responsible agent, but it was a climate change brought about by an impact.

The impact-caused winter was a major contribution, an important new idea. But then, in a great conceptual leap, Toon and his collaborators realized that a similar effect could occur today as a result of an all-out nuclear war, if sufficient dust were lofted into the high atmosphere. It was the birth of the idea of the "nuclear winter." Toon discussed these ideas with Carl Sagan, his former thesis adviser, who had analyzed the climate effects of dust clouds on Mars and had reached similar conclusions. They realized that the major contribution to the darkening of the atmosphere would not be dust lofted in the fireball, but soot from the extensive burning of cities that would very likely accompany an all-out nuclear war. This analysis was published in *Science* magazine, in a paper that soon became known as the TTAPS report, from the initials of the authors: Turco, Toon, Ackerman, Pollack, and Sagan.

The TTAPS report predicting a nuclear winter hit the entire defense community like a shock wave. A few years earlier, a committee of the

National Academy of Sciences, the most prestigious scientific organization in the United States, had analyzed the long-term effects of nuclear war and concluded that they might be relatively minor. Populations had sprung back from past disasters, such as the bubonic plague of 1350, with amazing resiliency and speed. But the Academy committee had never thought of the nuclear winter, which could be far more devastating than immediate blast or long-term radiation effects. Depending on the duration of the chill, it seemed possible that all higher forms of life could again be destroyed, this time in action brought on by ourselves.

Some critics of the TTAPS report said that the calculations had been overly simplified. The existence of the oceans, of storms, and of cloud formation had been neglected. The nuclear winter threat might not exist, or might be much less severe than the TTAPS report suggested. However, many sober-minded scientists felt that this criticism, although accurate, was irrelevant. The TTAPS group had shown that the best scientists in the world, in their previous thinking about nuclear war, might possibly have missed the most important, the most damaging effect. The TTAPS group might be wrong in their detailed climate calculations, but they were absolutely right in their demonstration that nuclear warfare was too large a departure from prior experience for us to have any confidence that we could predict its consequences. Nuclear winter might turn out to be right or wrong, but we had all learned a nuclear humility.

Soon after the Alvarez group published its discovery of the primordial impact, NASA commissioned a special study to figure out what to do if we discovered that an asteroid or comet was heading directly toward the Earth. One could not simply blow the asteroid to smithereens, since even a small smithereen can destroy a city. In 1908, a small piece of a comet, probably less than 100 feet across, hit the atmosphere above the Tunguska region of Siberia. It vaporized and never reached the ground, but the blast from the hot gases was equivalent to that of a 50-megaton nuclear explosion. Tens of thousands of square miles of trees were flattened. Observers flying over Tunguska a decade later reported that from the air the broken trees looked like matchsticks, with their bases all pointing back toward the place where the fragment would have hit.

The NASA study, which included Luis Alvarez as a participant, concluded that the best thing to do was to land on the asteroid, dig a hole in it, and set off a small nuclear explosion in the hole. In effect, the entire asteroid would become a rocket, with the blast of the explosion giving a thrust that would push the asteroid out of its old orbit. If you caught the asteroid early, a relatively gentle blast could deflect it enough so that it would miss the Earth. Physicist Roderick Hyde figured out that it would be even easier to deflect a comet, which is largely ice and dust. Detonate the nuclear weapon a few miles above the surface of the comet core, he suggested. The blast of heat would vaporize water on the surface of the comet, and this vapor would push the comet away. If it were sufficiently spread out, the explosion would not tear the comet apart.

We are the first creatures on the Earth capable of putting up this kind of "comet umbrella." It would be wise to practice on a few distant asteroids and comets, however, to make sure that the method works before we try it on a comet or asteroid that is headed directly toward the Earth.

3. Luie

I HAD originally gone to the University of California at Berkeley, in 1964, to become a nuclear physicist. I finished most of my course work within a year, but I knew I still had a lot to learn. The real problems to solve are not the ones at the end of each chapter in the textbook. The real problems come in dealing with the unexpected. The questions are vague and fuzzy. How do you pick a research topic? How long do you study a subject before you publish? What do you do when things don't look like you expected them to look, when your results surprise you? How do you recognize when to quit a not-completely-hopeless endeavor? Coping with these problems is the art of physics, and it is very similar to the art of business, or the art of art. You can learn it only from another physicist. So once you have finished your graduate courses, you are expected to apprentice yourself to one, your "thesis adviser."

The cutting edge of nuclear physics was a subfield called "elementary particle physics," the study of the "elementary" bundles of energy from which everything else in the Universe is constructed. In this field Berkeley was the center of the world. Almost every day some great scientist passed through Berkeley, pausing at the Lawrence Berkeley Laboratory to hear about the latest discovery. More rarely the scientist would tell about some new discovery made elsewhere. The vocabulary of the particle physicist was already colorful, with "strange particles" and the "eightfold way" being common lunchroom topics. It seemed as if a giant jigsaw puzzle had been laid out, and everyone was trying to fit the pieces together. But, as opposed to working a real jigsaw puzzle, in this case nobody knew what the picture would look like at the end, or what the boundaries of the puzzle were. Different groups specialized in different parts of the puzzle. It was a research area in which many people could work together, be they great scientists or only mediocre. Some people would fit more pieces together than others, and some were particularly good at finding that hard piece, the one that everyone else had given up on. But everybody was able to find some little way to help. Even though it was a fast-moving business, it looked to an outsider as if there was still a lot to do. There was bound to be a corner where even a graduate student could contribute. Maybe I could even be lucky and make a discovery.

One afternoon as I walked into an office shared by graduate-student teaching assistants (TAs), my friend Calvin Farwell was telling the other TAs about a new project just being started by Professor Luis Alvarez. Alvarez wanted to use cosmic rays, the intense radiation that comes from space, to study the properties of elementary particles. It was an extremely difficult and complex undertaking, for most of the radiation never reaches the surface of the Earth but is stopped by the blanket of the Earth's atmosphere. So, he explained, Alvarez was going to fly a complete elementary particle physics experiment from the bottom of a huge balloon, over 100 feet in diameter. Since there was no electrical power up at the top of the atmosphere, he planned to use superconducting magnets, a brand-new and untested technology. Neither spark chambers nor bubble chambers would work, and Alvarez had come up with a new kind of detector, something that combined spark chambers with nuclear emulsions in a clever way that made them far more effective together than they could have been separately.

I was entranced. Alvarez was famous as the man who had turned the bubble chamber into the primary discovery tool in elementary particle physics, but he was now abandoning it and inventing something new, something unheard of, to address new unsolved problems in an original way. The only question in my mind was whether such a great man would ever accept me as a student. I don't think I considered it a serious possibility. Instead of graduate students, Alvarez preferred having a dozen young Ph.D.s working for him, and these he treated as graduate students. But then I had a bit of luck. I became the head TA for Physics 4A in the new fall semester, and Alvarez was scheduled to teach it. So I would naturally meet the man and be able to make my own evaluation, without having to take the psychological risk of appearing to be soliciting a thesis adviser.

When I walked into the Physics 4A laboratory the next fall to have a meeting with the other TAs, there was a tall, blond, athletic-looking man wearing a sports coat sitting at one of the laboratory benches. He introduced himself with "I'm Luie Alvarez," avoiding the title "professor," which many other faculty members obviously favored. He said that he was there to help me organize the laboratory. I told him that he was welcome to attend our meeting but that I had an agenda already worked out. He promised not to interfere. He sat through the hour-long meeting, and listened patiently as I explained to the new TAs what their duties would be, how to prepare for the labs, and how best to guide the students. At the end of the meeting he came over to me and said that he clearly wasn't needed, so if it was okay with me, he would skip future TA meetings. I was both surprised and flattered.

I attended a few of his lectures to the huge class of about 300 students, and found his presentations very lively and interesting if not totally organized. Alvarez loved to sprinkle his lectures with personal stories from his experiences in the lab. All of physics seemed to have direct meaning to his own life and research experience. He didn't just practice physics, he lived physics.

I finally got my courage up, and after one class I asked him about his balloon project. We immediately sat down in the large Physical Sciences Lecture Hall. He wanted to know why I was interested. I said the project was *obviously* exciting. I didn't want to become a bubble-chamber physicist, or a particle-counter physicist, or a spark-chamber physicist, but just wanted to learn how to do experiments, experiments that had never been done before. It seemed to me that that was what his balloon project was all about. He told me I was absolutely right and suggested that I come up to the Radiation Lab right then to see the actual hardware and meet some of the people already working on it.

We drove up the hill on which the Lawrence Radiation Laboratory was located, along winding Cyclotron Road and through a forest of towering eucalyptus trees. This laboratory is devoted to basic research in science. Unlike the more famous Lawrence Radiation Laboratory located at Livermore, it does not engage in nuclear weapons design or other classified research. Alvarez didn't slow down as the guard waved him through the gate. The view from the hill was spectacular, with the entire San Francisco Bay area laid out below us. Two deer scampered off the road in front of our car. Alvarez explained that the deer had a refuge within the fence that divided the laboratory from the rest of the world. The road continued to wind, but buildings began to appear among the trees, steam rising from some of them. "That's the Bevatron," Alvarez said, pointing to a huge circular building with a flashing light on the door. It was the big accelerator that had been built by Ernest Lawrence, the Nobel laureate who had created the laboratory and had been Luis Alvarez's mentor. The Bevatron was used in most of the great discoveries made at the lab.

Finally we came to a long corrugated-metal building named, simply, Blg. 46. Inside were dozens of people in shirtsleeves working at benches and drafting tables and machines. Alvarez took me to a small area partitioned off from the rest, where a heavyset young man rose at our approach. He was William Humphrey, a physicist who was running the day-to-day activities of the project. To use a simile from the business world, Alvarez acted on the project like the chairman of the board, but had recently delegated the role of president to Humphrey. I was somewhat dazzled as Alvarez and Humphrey showed me around the rest of the building. I saw a difficult and complex project with many elements, each of which looked too complicated for me to understand, and a crew of several dozen physicists, engineers, machinists, and technicians all working together with a single goal, which had been defined by Alvarez. Alvarez always liked to take the lead in explaining the apparatus scattered around the building, but he always deferred to Humphrey for details. At the end of the tour, Alvarez asked me, "When can you start?" I had not understood much of what I had seen, and I didn't think I had asked any intelligent questions. In addition, Alvarez hadn't had a chance to check my grades or to look at my record in the Physics Department office, so he had no way of knowing whether I was a good student. But it was an offer I couldn't refuse.

Everyone in the Alvarez group called him Luie, except me. Alvarez had continued the tradition of informality that Ernest Lawrence had begun decades earlier. But I had been brought up to call all people older than myself by a title. Since I couldn't be the only one in the group to say "Professor Alvarez," I avoided calling him anything. It made it sound as if I had forgotten his name.

Alvarez's whole approach to physics was that of an entrepreneur, taking big risks by building large new projects in the hope of large rewards, although his pay was academic rather than financial. He had drawn around him a group of young physicists anxious to try out the exciting ideas he was proposing. Many of these people were later to leave basic research and become entrepreneurs themselves of one kind or another. The true rewards for this kind of work were in the challenge, in the adventure. Alvarez seemed to care less about the way the picture in the puzzle would look, when everything fit together, than about the fun of looking for pieces that fit. He loved nothing more than doing something that everybody else thought impossible. His designs were clever, and usually exploited some little-known principle that everyone else had forgotten.

I decided that my first job was to read and understand the long memo he had written on the balloon project. What was most striking about the memo was that Alvarez had no background in superconductivity, yet he was proposing to use state-of-the-art technology in this new experiment. But he had obviously done a lot of "homework," spoken to real experts, and convinced himself that there were no insurmountable obstacles to the new superconducting magnet. There were no known particle detectors that would work with sufficient precision in the confined space of a gondola hanging from a balloon. Alvarez had addressed this problem by inventing a new kind of particle detector, which combined the best features of some of the existing detectors, including spark chambers and nuclear emulsions, scintillation detectors, and a gas-filled Cerenkov detector. The concatenation of new technologies was overwhelming. It seemed to me that Alvarez must know everything there is to know in science.

I enjoyed the thought of working with things that I didn't really understand. Perhaps I had inherited this trait from my father, who loved to repair our TV set when I was a child. He had very little understanding of how the thing actually worked, but that never stopped him from fixing it. He opened the back, carefully defeated the interlock mechanism, and began sniffing around. Sometimes he would notice a tube wasn't lit. Other times he would find a capacitor (then called a "condenser") that smelled funny. In retrospect, I realize that what he was doing was the essence of good science. You look for something that smells funny, and you explore around it. You don't always understand what you are doing; if you do, it probably means that you aren't really near the forefront of knowledge. My dad was a detective, a solver of mysteries. He had a finely tuned nose, like Alvarez.

One day Alvarez noticed me sitting in an office reading, and asked what I was doing. I proudly told him how hard I was working on my "homework," his memo.

"You'll never learn experimental physics by sitting at a desk," he abruptly said. "Get over to Building 46, where the real physics is being done!"

I was embarrassed. "But I don't understand anything yet. I don't know how to help. I'd just be in the way," I protested.

I remember his answer very clearly. "That doesn't matter," he said. "Just go over there and hang around. Do anything anybody asks you to do. Sooner or later someone will see that you're there, and they'll ask you to hold a screwdriver. Get your hands dirty. Pretty soon you'll know how things are constructed. Once they have seen that you're around a lot, they may ask you to help test the apparatus. Before long you'll know how everything works. You can read memos anytime, in the evenings, at home, but you can only learn experimental physics by being in the laboratory, by doing it."

The next time Alvarez saw me was in Blg. 46, where I was helping test some spark chambers. We exchanged smiles. The part I had thought most difficult to learn, understanding the hardware, had turned out to be the easiest. But just as Alvarez had said, I couldn't learn it by reading memos or published papers. Even today, when I visit someone's laboratory and they sit me down in a nearby office to hear a lecture on their experiment, I protest. I want to go see the hardware, if possible to touch it.

As I have grown older, I realize that I am spending more and more time at a desk, planning new projects (as Alvarez had, in his memo) rather than building apparatus, rather than doing experiments. But even now a certain minimum contact with the hardware is absolutely necessary. In the Greek legend, the wrestler Antaeus was invincible as long as he touched the ground. Hercules defeated him by lifting Antaeus above his head, out of reach of the Earth, where he was able to squeeze him to death. This legend may have originated as a reminder to farmers to stay in close touch with the land, the source of their sustenance, and not to relegate all the manual work to hired laborers. The reminder applies equally well to experimental physicists, who must return periodically to their apparatus, to get their hands dirty, lest they forget why it often takes an hour to put in a screw.

A few weeks later I was helping another graduate student, Dennis Smith, mount a large photomultiplier tube in the Cerenkov detector, and I dropped it. It imploded just like a TV picture tube, with a loud and sickening crash. I had just destroyed $15,000 worth of hardware. Dennis consoled me. It could happen to anybody, he said. I thought it might be the end of my career. Fifteen thousand dollars was twice what I earned in a year as a research assistant. Some of the money could be recovered by firing me. A short time later I saw Alvarez, and confessed what had happened. "Grrrrr . . . reat!" he roared, and put out his hand as if to congratulate me. He shook my hand vigorously, but I protested. "Welcome to the club," he continued. "Now I know you're becoming an experimental physicist." To become a real member of the Mafia you have to murder someone. To become an experimental physicist, Alvarez seemed to feel, you had to destroy some expensive equipment. It was a rite of passage. "Don't do anything differently," he advised. "Keep it up."

To my awe of Alvarez was added that day a feeling of warmth. I overcame my shyness, and finally managed to address him as Luie.

After that Luie and I grew close. I became almost a scientific "son" to him. He shared not only his scientific ideas but also his thoughts on life and the world. I sensed that he truly appreciated my interest in learning from him, and I was very grateful for all the time he began to give me. It was still impossible to keep up with him in physics matters, to compete. I never seemed to have any idea that he hadn't thought of and dismissed years ago. It was hard on my ego. But then I did something that was one of the finest moments in my life. I made a truly great decision, one that determined my entire career. I said, quite consciously, almost aloud, "To hell with my ego!" I decided that the opportunity was simply too great. One of the great physicists of all time was aching to teach, and I would do everything I could to learn from this man. I would totally follow his suggestions for what projects to work on. If he had a new idea, I would drop whatever I was doing to work with him on it. My goal for the next several years would simply be to learn as much as I could about how Luie thought, how he approached physics, how he decided what to do, how he followed through. I would apprentice myself fully to Luie. I would not leave Berkeley until I felt that I was no longer learning.

What makes a good scientist? As the years began to pass, I noticed that the smartest students didn't always become the best researchers. That gave me some hope, since my course work was only slightly better than average. What was it they hadn't learned? I was awed by the careers of several of the faculty at Berkeley. There were more Nobel laureates at Berkeley than in the entire Soviet Union. And there was Luie, who had no Nobel Prize then but seemed to have more important discoveries to his credit than any of the scientists who had won the honor.

I made a mental list of Luie's great discoveries. He proved that cosmic rays were positive, probably protons; he discovered several rare isotopes; he discovered the radioactive process called "electron capture"; he found internal conversion in light nuclei; he discovered the magnetism of the neutron; he found the radioactivity of tritium; and he was responsible for a whole book of discoveries in elementary particle physics. To this list I could add several major inventions. He invented the triggering system for the atomic bomb (and flew in the chase plane over Hiroshima to measure the bomb's yield); he also invented ground-controlled approach (GCA), widely used for blind aircraft landings. But it wasn't the art of invention, but the art of discovery that I most wanted to learn. How did he do it? What secret knowledge enabled him to make discovery after discovery? If only I could identify what it was, then I could try to learn it.

Skepticism, the ability not to be fooled, was clearly important, but it is also cheap. It is easy to disbelieve everything, and some scientists seemed to take this approach. Sometimes Luie was skeptical, but more often he seemed to embrace crazy ideas, at least at first. He rarely dismissed anything out of hand, no matter how absurd, until he had examined it closely. But then one tiny flaw, solidly established, was enough to kill it. His openness to wild ideas was balanced by his firmness in dismissing those that were flawed. He had a finely honed skepticism. Perhaps that was part of his secret talent.

Scientific training doesn't keep your senses from fooling you, but a good scientist doesn't accept the impressions his senses deliver. He uses them as a starting point, and then he checks, and double checks. He looks for additional evidence, and for consistency among his measurements. A scientist differs from other people in that he *knows* how easily he is fooled, and he goes through procedures to compensate.

Luie had learned about the art of discovery from his father, Walter Alvarez, a physician and medical researcher, famous for his newspaper medical column "Ask Dr. Alvarez." Some people also gave him credit for restoring the prestige of general practitioners at a time when specialists had been getting all the attention. His son had obviously learned at least one great lesson from his father, for Luie had become a general practitioner of physics. The senior Alvarez believed the key to discovery was not being "lazy." He had come very close to winning a Nobel Prize in medicine. He had had the idea of treating pernicious anemia with liver, but he had never really followed up on his observations. He had been too lazy, or perhaps too skeptical. The hardest discoveries to take seriously are often your own. A few years later, George Whipple and others pursued the liver-treatment idea, developed it into the standard method of curing the disease, and were awarded the Nobel Prize in medicine. Walter Alvarez never came that close to a Nobel Prize again, but he made sure that his son learned the lesson.

Luie had no Nobel Prize, but he believed that he had narrowly missed winning it at least twice. He had almost discovered fission, the "transmutation of the elements," when he was bombarding uranium with neutrons. He would have found it if he had run his apparatus for a half hour instead of just a few minutes. When he read in a newspaper that fission had been discovered by Otto Hahn in Germany, he was immediately able to confirm the discovery, but it was too late. It was not his discovery.

The second Nobel Prize he missed was the discovery of secondary neutrons from fission, the process that makes the chain reaction possible. Luie had produced neutrons with the cyclotron, and bombarded various materials with them. He had all the apparatus needed to see that sometimes when a neutron hits a uranium nucleus, two neutrons come out, but he hadn't looked long enough. One neutron can lead to two, two to four, four to eight, and so on; about seventy-five generations later virtually every atom in the sample has emitted a neutron, unless the tremendous heat generated has blown the sample apart. The chain reaction is the basis for the nuclear reactor and the atomic bomb.

The art of physics consists in knowing what to work on and for how long. Many physicists have wasted their careers following up on uninteresting discoveries. But to me, as a graduate student awed by Luie's successes, it was not easy to discern a pattern behind his choice of projects. His balloon project was clearly exciting, although perhaps *too* difficult to lead anywhere. Soon after I joined this experiment, Luie began the pyramid project. It looked like fun, but it wasn't the sort of project I had expected a world-class physicist to work on.

Luie had read about the pyramids as a child, much as I had read about the dinosaurs. He had read of how Howard Carter had found the tomb of Tutankhamen by a combination of meticulous planning and preparation mixed with a good deal of daring. From his readings Luie was convinced that there were other unfound chambers, probably in the pyramids at Giza themselves. If success means to have a dream of childhood come true, then Luie's crowning success would be his discovery of a hidden chamber full of gold, treasures, and history. And in 1966, Luie finally thought he had invented a method of looking inside the pyramids, tantamount to x-raying them. His idea was to apply the methods of nuclear physics to archeology, to use a special kind of penetrating natural radioactivity (muons) in place of x-rays.

Luie worked out virtually all the details before he even told anybody what he was thinking of, and he put these in a memo. Real x-rays wouldn't penetrate deeply enough into rock, so Luie planned to use muons, elementary particles created by cosmic radiation from space, that can penetrate hundreds of meters of rock. To estimate the flux of muons, Luie used a rule of thumb: roughly 1 cosmic-ray muon passes through your thumb every 10 seconds as part of the natural "organic" radiation that always surrounds (and penetrates) us on Earth. All the details were worked out before any apparatus was built. It was experimental physics from a desk. Many scientists think that experimentalists don't think; they work with hardware. Luie believed that you had to know and understand a subject thoroughly before attempting a measurement.

I was beginning to see, in the way Luie worked, a possibility for my own research. There were always others far more talented in mathematics than me, and there were always others far more scholarly and comprehensively knowledgeable. I had found physics frustrating because there was too much to learn. But Luie's approach to physics wasn't mathematical or comprehensive; it was clever and inventive. He had learned just enough about every subject; he could go back and fill in the gaps later, when and if that was necessary. The gaps in his knowledge were surprisingly large, but not detrimental to his work. Even though he knew little about quantum mechanics, he had discovered internal conversion in light nuclei, an extremely important and unexpected quantum-mechanical process. His knowledge of theoretical optics was extremely spotty, yet he had used diffractive optics in a clever and original way to invent a method of landing airplanes during conditions of zero visibility. (For the invention of this ground-controlled approach during World War II, Luie had won many awards and been thanked by hundreds of pilots who owed their lives to him.) He seemed to have a knack for learning just the right amount about everything, and for spending the time he saved inventing and bringing together ideas from disparate fields.

Luie's pyramid project was an adventure, science being used for exploration. He believed that he had inherited the legacy of the great explorers. He had read and reread the journals of Sir Richard Burton, the first non-Muslim to enter Mecca, and of Captain James Cook, explorer of the Pacific. Explorers must be prepared for the unexpected. They must be ready to allow fate to lead them in new directions. They have to be broadly educated and aware of where the true frontiers are. Science is the modern-day tool to explore the world. The real excitement in science is the excitement of discovery, particularly the discovery of something you had no reason to expect was there.

Luie claimed to be driven by a sense of curiosity, but if that were true he could have spent his time becoming a scholar. With so much physics to learn, why work hard trying to discover a few small facts in physics that nobody else knows? The truly curious don't waste their time fitting together jigsaw puzzles, since there were so many beautiful pictures to be seen in books and museums. Does anybody really solve jigsaw puzzles in order to see the picture? No, they do it because of the fun of solving the myriad of little puzzles along the way. Luie was a puzzle solver, an adventurer, an explorer.

Of course, not all explorations succeed. Ponce de León never did find the Fountain of Youth. Luie found no chambers in the pyramid. Some newspapers mistakenly reported that he hadn't found *any* chambers, and Luie was quick to point out their error. He did more than search without success; he had searched and found that there were *no* chambers inside. The problem with risk-taking is that sometimes you fail. Resiliency is not a virtue for research in physics; it is an absolute necessity.

One morning in 1968 I was having trouble tuning the radio to the morning news. What I heard was "garble . . . garble . . . Nobel Prize in chemistry . . . garble . . . garble . . . physics . . . garble . . . garble . . . Alvarez . . . garble. . . ." I ran into the next room looking for my wife, Rosemary, shouting, "Luie won the Nobel Prize! Luie won the Nobel Prize!" I lifted her up in the air and spun her around, surprising myself with my own strength. We stopped dancing for a moment as I suddenly wondered if I had misheard. It made so much sense to give him the prize, but I thought Luie had missed his chance when the award had been given to Donald Glaser several years earlier for the invention of the bubble chamber, the device that Luie had perfected and used for myriad discoveries of elementary particles. All my joy was based on one word, which I might not have heard correctly. I called up the lab, and got Luie's secretary on the phone. "Ann, I just heard on the radio . . ." "Yes," she interrupted, "isn't it wonderful!" I rushed to the lab. Everyone was celebrating, waiting for Luie to show up. He had been called at home, early that morning, by a newspaper reporter. The prize was given for all the discoveries that had come from the bubble chamber. Luie was the sole physics winner that year.

There was cheering down the hall, and I guessed that Luie had finally arrived. Everybody was deliriously happy. For someone with a reputation among outsiders as a tough curmudgeon, Luie had obviously generated a great deal of loyalty and love among those who had worked with him. I was even surprised at my own happiness. It was the most exciting day of my life.

4. Dinosaurs

IN 1976, seven years had passed since I had received my doctorate. I had stayed at Berkeley, and had followed Luie's model in pursuing several projects of my own. Two of these had become particularly successful: a measurement of cosmic microwave radiation, and the invention of a new method for radioisotope dating. I would soon be rewarded with a faculty position in the physics department at Berkeley and two prestigious national awards. A friend advised me that it was easier to get a good reputation than to keep it. I was looking for something new to do.

I received a telephone call from Walter Alvarez, Luie's son, who was then working at Columbia University's Lamont-Doherty Geological Observatory. I had never met Walter, and this was only the second time we had spoken on the telephone. He had heard a theory that the dinosaurs had been killed by a supernova, and he thought that he might be able to prove or disprove the theory by making some measurements of a thin clay layer he had found in Italy. He said he needed my help, since he had learned from his father that I had invented a sensitive radioisotope-dating technique. His ideas were flowing so quickly over the phone that I couldn't keep up. I finally slowed his pace by saying, "I thought that the dinosaurs had been killed off by smarter mammals." As soon as I'd made this comment, he realized the depth of my ignorance, and he transformed himself from an excited scientist into a patient professor. He said, "Let me tell you a little bit about the dinosaurs." I was grateful for both the change of pace and the absence of condescension in his tone.

Most of what I knew about dinosaurs I had learned as a child, from the one book the local public library had on the subject. (In the olden days, unlike today, we were not inundated with dinosaur books, dinosaur models, and stuffed, furry, cuddly dinosaur dolls.) I knew the dinosaurs had mysteriously disappeared at the end of the Cretaceous period, 65 million years ago. Most people believed that their bodies had simply grown too big for their brains, and they had become too dumb to compete with the mammals; the word *dinosaur* has even become a derogatory metaphor for anything grown too big for its own good. This is what I thought I knew about dinosaurs. But as Josh Billings, the nineteenth-century humorist, once said, "The problem with most people isn't so much their ignorance; it's what they know that ain't so."

Walter quickly corrected my misconceptions. He explained that 65 million years ago, the brightest of the mammals were no smarter than the most intelligent of the dinosaurs. The mammals had been no recent invention of Nature; they had competed unsuccessfully with the dinosaurs for 100 million years. Moreover, most of the mammals that lived back then became extinct at the same time. The popular myth that the dinosaurs had been killed by clever little mammals with a taste for dinosaur eggs was no more worthy of respect than the facetious claim that the dinosaurs had died because they were too big to fit on Noah's Ark.

Whatever killed the large lizards did not particularly specialize in dinosaurs. Over two thirds of all species living at the time disappeared along with them, including not only mammals, but also fishes, corals, shellfish, and even the microscopic single-celled animals known as forams. So much of the biomass was destroyed that paleontologists called the catastrophe a mass extinction.

At that time the most reputable theory for the extinctions attributed them to climate change. This in turn may have been brought about by continental drift, also known as plate tectonics because it is not just the continents that move, but larger pieces of the Earth's crust, known as plates. North America and Europe are presently moving apart at a speed of about 1 millimeter per month as new seafloor is created at the mid-Atlantic ridge. "That's about the same rate that your fingernails grow," Walter said. A millimeter per month adds up to about a centimeter per year, a meter every 100 years, 10 kilometers every million years, and 10,000 kilometers every billion years, comparable to the size of the

Earth. As these plates move, many other things happen. The sea level can rise or fall, the configuration of the oceans can be altered, and the climate, worldwide, can change. About 65 million years ago such motion caused the great inland seas of North America to dry up, a change sufficient by itself to trigger an ecological catastrophe. But Walter hadn't found the reputable theories of mass extinction convincing. He was particularly bothered by the abrupt disappearance of those microscopic animals known as forams, short for foraminfera. What could simultaneously kill both the giant dinosaurs and the microscopic forams, and do it all over the globe?

Malvin Ruderman, a professor of physics at Columbia University, had proposed that the creatures were killed by a nearby exploding star, a supernova. In such an explosion the outer layers of a star blow off, and for a few seconds the power poured into space is greater than the combined power of all the other stars in the Universe. For a few weeks the expanding plasma glows with the brightness of 10 billion stars, greater than that of all the other stars in the galaxy put together. Ruderman speculated that intense x-rays from such an explosion destroyed the ozone shield of the upper atmosphere and allowed life-killing ultraviolet radiation from the sun to reach the surface of the Earth. According to this supernova theory, the disaster lasted only a decade, the time it took the atmosphere to restore the shield. Walter noted that this period was much shorter than the millions of years required by other theories, and so it was the duration of the catastrophe that he was most interested in measuring.

Sedimentary rock forms on the seabed as coccoliths, the remains of shells and skeletons of small sea animals, drift to the bottom and are gradually compressed by the weight of further material falling from above. Walter had been working for several years at various sites in Italy, where he had painstakingly identified all the geologic stages in the rock, and he knew exactly where to find the rock formed at the time the dinosaurs were killed. Just above the rock formed in the Cretaceous period, the last age of the dinosaurs, the continuous limestone was interrupted by a layer of clay less than a half-inch thick. Below the clay layer were abundant forams, but above it they were virtually absent. Walter speculated that the clay was laid down during the great catastrophe, when there were few sea creatures to form limestone, but clay continued to wash down from continental erosion. If we could tell how long it took for the clay to be deposited, we would know how long the catastrophe had lasted.

Walter did not have to twist my arm to get me interested in helping. The measurement he wanted looked tough, but possible. I agreed to visit him in New York to work out some plans in detail. While there I would also give a colloquium about the cyclotron method and its applications to geology. There were many experts at Lamont who could criticize our plans, and thus help us anticipate problems.

The death of the dinosaurs had been the first unsolved mystery of science I had learned of as a child. After a while I stopped thinking about the difficulties ahead and, instead, indulged myself in the fantasy of solving the problem. It doesn't hurt to fantasize a little, and I had better do it while I could. Luie had taught me that most clever ideas are proven to be wrong within a few days. If you manage to have one good idea every week, then you can expect to find one every few months that is really worth pursuing. Every few years you might have an idea that could lead to a discovery. The important thing is to keep working, thinking, trying out crazy ideas. If you can keep up the pace, then maybe once in your lifetime you might stumble on something really important, something that could change the course of science.

As I drove up to the Lamont-Doherty Geological Observatory, I pondered the odd title given to this laboratory. To me, *observatory* meant telescopes, chilly nights spent on mountains trying to work in virtual darkness. Why had the geologists borrowed this term from the astronomers? True, neither astronomers nor geologists can do experiments, at least not in the sense that physicists do. The stars and the rocks just sit out there, giving at most a snapshot of what the Universe is like right now. You can't change them in any significant way; you can only make detailed, careful observations of their present configuration, and try to deduce what they must have been like millions or billions of years ago. Astronomers and geologists are observers, not experimenters, observers of the experiments done by Nature. So perhaps "geological observatory" *was* just right.

When I had talked to Walter on the phone I had imagined he was in some dingy office that perhaps looked out on other buildings, but in fact his office was in an elegant old mansion that had once served as the manor house on an estate. From the site high over the river, one could see the high cliffs of the Palisades and the rolling hills of the Hudson Valley. Geologists certainly know how to live.

For some reason it had never occurred to me that Walter would look like his father. But he did. There he was, tall and lean, with the characteristic smile and blond hair of my former thesis adviser. I have had similar experiences many times, yet it always surprises me to see the clear effects of heredity in people. Perhaps I subconsciously like to forget that I am a biological as well as a cultural creature. Walter looked like a man who had spent a good deal of his life outdoors, and he had a strong handshake. But his voice was somewhat gentler than that of his father. It had a slight twang to it that reminded me of Henry Fonda.

My colloquium at Lamont was scheduled early in the day. About a dozen geologists came to hear me talk about "accelerator mass spectrometry" and how it might be used for dating sedimentary rock. The method used a particle accelerator such as a cyclotron, commonly known as an atom smasher, to accelerate a few of the atoms in the rock to a speed near that of light. When the atoms move with this velocity, special detectors can be used to identify and count individual atoms. One particular atom that looked useful for geology was Be-10, an isotope of beryllium with a total of 10 protons and neutrons. It is a radioactive atom, created when the cosmic radiation from space hits the atoms of oxygen and nitrogen in the atmosphere, breaking them into smaller pieces. Most of this Be-10 gradually falls to the surface of the Earth or into the oceans. As it settles to the seabed it gets trapped in whatever rock happens to be forming. We believe that this rain of Be-10 on the Earth is constant, since we think that the influx of cosmic radiation is constant. So if the rate of rock formation is changing, the density of entrapped Be-10 will change along with it. If a lot of rock forms in any particular year, then the density of Beryllium-10 will be low. If the rock forms slowly, the density of Be-10 will be high. By measuring the density of Be-10 atoms entrapped in the sample, we could answer Walter's question about the rate of formation of the mysterious clay layer.

Be-10 is radioactive, and the data books gave its half-life as only 2.5 million years. That means that in each 2.5-million-year interval after it is laid down half of the remaining Be-10 in the layer will disappear. In 65 million years there are 26 such intervals, so the fraction of Be-10 left after that time can be calculated as $(1/2) \times (1/2) \times (1/2) \times (1/2) \times \ldots$ with the term "$(1/2)$" appearing 26 times. A few quick button pushes on a pocket calculator had given Walter and me the answer during our phone conversation: the original Be-10 had been reduced by a factor of 67 million. Virtually all of it was gone. Nevertheless, the new cyclotron method was so sensitive that it might be able to detect even this tiny remnant. The measurement looked difficult, but still possible.

The geologists in the audience asked pertinent questions, and raised many objections. I was able to handle the physics, but I had to defer to Walter for virtually everything relating to geology. There were clearly many subtleties in this business that I wasn't even aware of. By the end of my presentation I had the strong impression that Walter knew my subject much better than I knew his.

Afterward Walter took me to see the enormous "library" of seafloor cores that were stored at Lamont. These were samples of rocks taken by the famous oceanographic ship *Glomar Challenger*, which had spent years (and many millions of dollars) drilling sample holes in the bottom of the oceans all over the world. This ship was similar to, but should not be confused with, the *Glomar Explorer*, which achieved notoriety by being secretly run by the CIA, and which reputedly recovered parts of a Soviet submarine off the seafloor. It is ironic that the *Glomar Challenger* was really an explorer, and the *Glomar Explorer* was used, in effect, to *challenge* the Soviets.

The huge core library was located in the basement of one of the larger buildings. Neatly laid out in rows over a hundred feet long were thousands of metal trays containing cylinders of rock, each one carefully marked with a notation similar to that used in the Dewey decimal system. The age of most of this rock had never been determined, but here it was, a sample of the world, sitting at Lamont waiting to be measured. It was invaluable and irreplaceable (except at great cost), full of many secrets, waiting for clever scientists to decode it. The history of the world was there, if only we could decipher it. Much progress had already been made by using fossils and remnant magnetism to identify the layers. Now Be-10 and the accelerators might help.

Walter didn't like to stay indoors, and we talked while walking around the beautiful grounds. He delighted in pointing out where the servants' quarters had once been, where the horses had been kept, which buildings had been used by the master of the estate for entertainment. It was characteristic of him, as I more fully appreciated when I visited him in Italy many years later, that he loved to learn the history of any place where he worked. We talked about geology and physics, and I continued to be impressed with his quick grasp of physics ideas. He was better at this than most physicists I knew, and yet he was very modest and aware of the limitations of his knowledge.

When I told him about the work I had done with his father, he became particularly attentive. He wanted to know everything I could tell him about his father. I told him how I had apprenticed myself to Luie when I was a graduate student, and how I had once pledged myself to learn everything he was willing to teach me. The apprenticeship was still continuing; after a decade I still had a lot to learn.

But I became self-conscious when I realized that I seemed to know Luis Alvarez far better than Walter did. His parents' marriage had broken up just as he was finishing high school, but if he had ever harbored resentment over their divorce, he had clearly left it far behind now. He had childhood memories of his father, but knew almost nothing about his father's professional life. And in the interval he had become a scientist himself. Thus he was consumed with curiosity about his father's work. He wanted to know every detail of his father's methods, how he worked, what he was like to work with, how he thought. I surprised myself by knowing the answers to all his questions. In some sense, because I felt that I had become a surrogate son to Luie, I found it awkward talking to Walter, the real son. But my embarrassment faded as I realized that Walter harbored no jealousy, only great curiosity. I soon found myself talking to him freely. I loved telling him about the excitement of being exposed to Luie's constant barrage of new ideas.

When I flew back to Berkeley, I thought of Walter as a new friend, potentially as a long-lost brother, and I hoped that our collaboration would grow. But the measurement he wanted wouldn't be easy. The new method, if everything worked perfectly, would barely reach back 65 million years. Perhaps years of further development would be necessary to give the required sensitivity.

Before visiting Walter, I had given a talk at the Research Progress Meeting (or RPM) at Berkeley about the cyclotron method. My work in this field had begun when Luie had had the ingenious idea of converting a large cyclotron into a mass spectrometer, and then using it to search for quarks. With the help of William Holley and Edward Stephenson, we had succeeded in making the search, but after two years of effort we had found none. It had turned into a "null" experiment, an experiment that states with great precision that something is not there. Null experiments are the least exciting kind done by scientists. Luis' project to x-ray the pyramids had likewise proven to be a null experiment.

At the Research Progress Meeting I had reported on our null quark search, and on my idea of applying the method to the detection of isotopes such as C-14 (radiocarbon) and Be-10. Grant Raisbeck, a young physicist with a background in nuclear science, had been one of those in the audience. As he told me later, he spent most of the time while I was speaking "kicking himself" for not having invented the cyclotron method himself. He had been interested in Be-10 for many years, he said, and had realized its potential for geology, but he knew that the standard methods for its detection were very difficult and insensitive. He decided that I had found the key to unlock the problem. The accelerator method would open up a new field of research. Shortly after my talk, Raisbeck devoted himself fully to the exploitation of the new method. In a few years he became the world's expert in the use of the technique for Be-10, and made hundreds of measurements, using several accelerators in France, where he continued his research.

In July 1976, after I returned from my visit with Walter Alvarez, I found Raisbeck waiting in my office. I was pleased to see him again, particularly because I had the most exciting Be-10 experiment in the world to tell him about. Walter and I were going to try to solve the mystery of the dinosaurs' extinction. Raisbeck found this application interesting, but complained that I was using the wrong value for the half-life of Be-10. I showed him the number in the table of the isotopes: 2.5 million years.

"I'm sorry," he said. "That number is wrong. Don't you remember? I told you after you gave the RPM. I measured the lifetime myself. It is 1.5 million years, not 2.5."

I vaguely remembered Grant coming up to me after my RPM, but at that time I hadn't been very interested in Be-10. I felt that he couldn't be right about the half-life. The value of 2.5 had been obtained by Nobel laureate Edwin McMillan, recently retired as director of the Lawrence Radiation Laboratory. I was certain that McMillan was too good a scientist to have published such an incorrect number. But Raisbeck was equally certain that McMillan was wrong.

If Raisbeck was right and McMillan wrong, it would end the proposed dinosaur study with Walter. Sixty-five million years is only 26 half-lives if the value is 2.5 million years. But if the value is 1.5 million years, it would be over 43 half-lives. The additional 17 half-lives would reduce our signal by a factor of $(1/2)$ multiplied by itself 17 times, that is, $1/131,000$. The Be-10 signal would be 131,000 times weaker than we had planned. A difficult experiment would turn into an impossible one.

I wasn't sure what to do, and I told Luie about the discrepancy. He immediately called McMillan, who didn't take long to discover what had happened. He located his original notebooks and compared his results with those in his published paper. He found that the value in his notes was closer to Raisbeck's "new" value of 1.5 million years than to his own published value. He traced the discrepancy to a sentence in his paper in which he calculated the *mean life*, a number 1.4 times larger than the half-life. To his amazement, he had left out the factor of 1.4 (1 divided by the natural logarithm of 2) and accidentally published the mean life instead of the half-life. It was only a transcription error!

McMillan chided me for having used the value given in the table of the isotopes without ever looking back at his original paper. "Sloppy physics," he said. "Physicists today are lazy. Nobody reads the literature any more. If you had read my original paper carefully you would have caught the error."

He was right, in part. I had done a lot of work, planned a lot of research, without checking one critical number. Had I read McMillan's article I might have noticed the discrepancy. But maybe not. It's very hard to catch mistakes like that. Nobody reads the literature any more, I nastily thought, including the author. The dinosaur project was dead, dead as a project could get. Killed by a misprint.

I called Walter and told him the tragic news. Too bad. But that is the way projects go. Very few ideas really lead to an important discovery. You mustn't sulk every time one of your ideas fails. Just keep on having ideas. Once per week. Keep at it. Don't stop thinking.

5. Iridium

MY FANTASY of solving the riddle of the dinosaurs was dead, and I was relieved. My scientific life was already pretty full, and the failure of this project made room for other things. I rationalized that we never would have found anything important anyway. I was better off not wasting all that time. In April 1977, a paper of mine on accelerator mass spectrometry was published in *Science*, complete with the first radio-isotope-dating measurements made with the Berkeley cyclotron. It attracted the interest of a diverse and fascinating collection of scientists. Another project I had started, one to detect the velocity of the Earth with respect to the rest of the Universe, was beginning to show exciting results. We found that the Earth was moving at just over 1 million miles per hour with respect to the distant galaxies. It soon proved to be an important number in cosmology, in helping to understand the large dynamics of the Universe. With these two achievements in hand, I was sitting on top of the world, and it was easy to forget about dinosaurs.

In the meantime, Walter Alvarez was trying to decide whether to come to Berkeley as an assistant professor of geology and geophysics. The move would mean lower rank for him, and he would have to break one of the cardinal rules of the ambitious: Never accept a salary cut. He had never been greedy for money, however. He saw research opportunities at Berkeley that didn't exist at Lamont. He loved the academic atmosphere of a university campus and the diversity of exciting research being done. And I think he was also drawn by his father, his fascinating father, whom he knew so little.

Luis Alvarez freely admitted that there was nothing he would rather do than write a paper with his son. He had always been somewhat baffled by Walt's interest in geology and in its little problems: how did *that* mountain form, or *that* valley? Where were the *big* problems, the problems of universal importance? The hardest thing for any scientist to learn is how to choose the best projects to work on. Perhaps Walt could benefit from a little guidance. Walt, like me, seemed to be interested in learning as much as he could from Luie. Maybe he could even get his father interested in geology. I drifted away from Luie into my own world as Walt grew much closer to him.

Walt decided to spend a year at Berkeley in 1977–78 as a way of testing the water. When he arrived, he gave his father a special present, a rock that he had carefully cut from an exposed section near Gubbio, Italy. The rock has several layers, characteristic of its sedimentary origin. He had encased the rock in plastic and polished one side of it. Walt had Luie look closely at three layers in the rock with a pocket magnifier. The lowest layer was a bleached limestone, calcium carbonate left from an uncountable number of microscopic coccoliths. Embedded in this limestone were the larger circular objects called "forams." The fossils from these one-celled animals were very abundant in the layer, and quite diverse. Just above the bleached limestone was a thin dark layer about half an inch thick. This clay layer had no fossils in it. Above the clay was more limestone, somewhat darker than the lower layer. Walt asked Luie to look there for forams. He couldn't find any. Only one species of forams had survived, and its members were too small to be seen with the magnifier. The numerous ocean forams alive today all evolved from that one tiny species that somehow made it through the catastrophe. The other forams had vanished, Walt told his father, at the same time the dinosaurs had disappeared. The rock sample was from the same layers that Walt had wanted me to analyze with the cyclotron.

Luie later described this experience of looking at Walt's rock as one of the most exciting moments of his life. Even though he had heard Walt describe the sweeping nature of the catastrophe before, his reaction was different now that he held that rock in his hand, now that he had seen the forams with the hand lens. It was not an abstract problem any more. Huge dinosaurs and tiny forams destroyed at the same time. How could it have

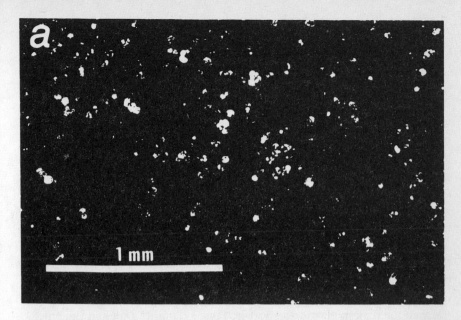

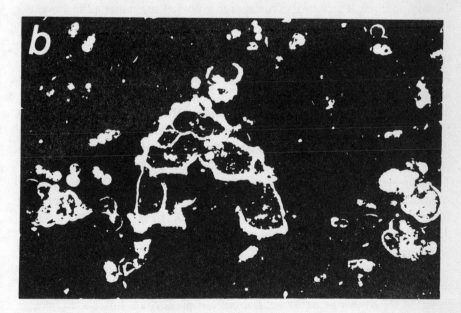

Microphotographs of the forams in the rock just above and below the clay layer. Many large forams in the lower layer are all gone in the upper layer, killed by the impact. Only one species of small foram is left.

happened? It was hard to imagine that any of the standard explanations for the catastrophe really accounted for this. Walt's rock could not be ignored. It was a *big* question in geology. Something dramatic had happened, something that had affected our own existence, perhaps made it possible.

The clue was that thin layer of clay. Luie had become interested, fascinated. The cyclotron method had proven inadequate, but such failures didn't deter Luie. There must be some physical way of determining sedimentation rate, of determining whether the clay had been set down in 100 years or in 100,000 years. If the problem is sufficiently important, then there *must* be a way. It was an odd project for a nuclear physicist to choose. Luie had not had much luck in picking research projects in the previous few years, and yet his earlier career made me think of him as an expert in the art of discovery, of finding the unexpected. Now he had decided to look at his son's clay layer, and determine how long it had taken to be laid down.

Although I had been excited about the project when I thought my accelerator method was applicable, now I doubted Luie's judgment. As Luie's student, I had studied his approach to picking research topics. I had decided that the value of a project was given by a formula that looked something like this:

$$\text{Value} = \frac{(\text{Probability of success}) \times (\text{Importance})}{\text{Effort}}$$

Although the importance of understanding the clay layer was high, the probability of success was low, and the effort required large. He had made so many discoveries that he no longer seemed interested in making new ones, unless they had the potential of being truly revolutionary. He was taking big risks, but he was consistently losing. The balloon project of my graduate-student days had led nowhere important. The pyramid project had found no chambers. Luie's judgment was failing, I decided.

When I tried to apply the cyclotron to the problem, I had used a technique that I already had in hand. The effort required was small, boosting the value in the equation. Luie had no relevant technique. He would have to invent something new. He would be wasting his time. It would be another project to add to his list of null experiments. Maybe he was really just interested in working with his son. That was probably it. Good for him. He had recently retired. It was silly to expect him to continue working at the intensity he had until now. No, I wasn't jealous, I thought. I had enough of my own projects to work on.

While I was rationalizing, Luie was thinking about physics. He couldn't get the clay layer out of his mind. He played with the encased sample Walt had given him. There must be *some* method of measuring the sedimentation rate, and he felt confident that he could find it. He drew on his vast knowledge of nuclear physics. If there was a method waiting to be discovered, he knew that it would probably be in this realm. He had learned this subject when he was a young scientist working for Ernest Lawrence, when Luie was in charge of updating the isotope tables. Back then, discoveries seemed to come in every week, and he knew each isotope of each element as if it were a good friend. He knew countless miscellaneous facts about the isotopes, too. Could any of this knowledge be exploited?

Which atoms in the clay layer might be used to determine sedimentation rate? Beryllium-10 from the cosmic rays had proven too weak a signal. Was there anything else that came from the cosmic rays? What about the micrometeorites that constantly bombard the Earth? Luie knew that dust was settling on the Earth from the constant rain of meteors that vaporize when they hit the Earth's atmosphere. Only the larger meteors glow brightly enough for us to see at night. Far more numerous than the meteorites are the micrometeorites, smaller than sand grains. Airplanes flying at high altitude to sample air with paper filters had picked up a fine dust that turned out to be the residue of micrometeorites that had lost their tremendous speed (typically 20 miles per second) in the very high atmosphere and were drifting slowly down. Luie realized that micrometeorites might be part of the solution to Walt's problem. If the rain of micrometeorites is constant, then one could determine the sedimentation rate by seeing how many micrometeorites had become trapped in the new rock as it formed under the oceans.

How could one find and count the micrometeorites? It could prove to be a hopelessly difficult task if done manually. Much of the material was too small to be seen even with a microscope. Luie's mind returned to his special knowledge of nuclear physics. The composition of the meteorites was known to be slightly different from the composition of the Earth's crust. Was it sufficiently different that the difference could be found chemically? He disappeared into the library at the recently renamed Lawrence Berkeley Laboratory to get the accurate numbers. He checked the chemical composition of both meteorites and the crust of the Earth. He wasn't satisfied with summary tables, with their potential for errors. He found and read the original references. And he found something potentially interesting.

Platinum, gold, and all the elements in the same group on the periodic table are 10,000 times as abundant in meteorites as in the Earth's crust. How can this be, considering that the Earth and meteors have a common origin in the dust of the early solar system? The answer is simple: The platinum-group elements alloy readily with iron. (In fact, platinum and gold are called "siderophile" elements, which translates to "iron lovers.") The early Earth was very hot, its rock melted by the heat generated by natural radioactivity in the rock and the energy contributed by numerous meteor impacts. Subsequently, the Earth cooled and developed a solid crust, but only because most of this radioactivity had decayed away. Even so, most of the radioactivity we absorb in our bodies during our lifetime still comes from rock and soil, not from cosmic rays or x-rays. The Earth was large enough so that its own gravitation was strong enough to pull most of the dense liquid iron down to its core. This separation of the iron from the crustal material is called "differentiation." In contrast, most meteorites, as they floated in space, were too small for differentiation to take place.

As the iron sank to the Earth's core, it pulled the siderophile elements, gold and platinum, with it. Ironically (no pun intended), the gold and platinum were swept down much more completely than the iron. Although much iron remained behind, most of the siderophile elements vanished deeply into the Earth. Thus the siderophile elements are not particularly abundant in the meteorites, but they are virtually absent in the crust of the Earth.

In an astrophysics handbook, Luie found that approximately 400 tons of meteorites hit the Earth every day. Of this, about $1/2$ part per million is iridium, or about 200 grams per day, or about 70 kilograms per year. The area of the Earth is about 5×10^{18} cm^2 (5 followed by 18 zeros). Limestone forms at the rate of about 1 micron, a millionth of a meter, every year. Putting these numbers together, he found that the meteoritic component in the limestone would be at the level of about 5 parts per billion.

That was what he had hoped. Meteoritic dust would inject more gold and platinum into the sedimentary rock than would any other source. If he could detect these elements in the rock, their concentration would be a direct measure of the sedimentation rate. How could they be detected? He didn't need to know the chemical composition, just which elements were present. No need to look at molecules, just atoms. To identify atoms it is sufficient to identify their nuclei. Nuclei are often easier to identify than atoms, because their signatures are not affected by the details of what molecules the atoms happen to be in. Neutron activation analysis was one of the most sensitive methods for detecting nuclei of rare elements.

In neutron activation analysis, the material (sedimentary rock, in this case) is exposed to neutrons from a nuclear reactor. Neutrons do something special that most other forms of radiation can't: They make the material exposed to them radioactive. There are fewer than ninety-two elements found in Nature, and the radioactivity of many is quite distinctive. If you are interested in measuring the amount of platinum, and don't care what its crystal or chemical form is, you can expose your sample to neutrons and look for the characteristic radiation from neutron-activated radioactive platinum.

Luie checked to see which of the siderophile elements would be best to use. Those elements that were the least chemically reactive, the "noble elements," were the most likely still to be in the clay layer after 65 million years. These were gold, platinum, iridium, osmium, and rhenium. It was important that the element have a high probability of absorbing a neutron. This probability varies from isotope to isotope, and Luie checked each one in the table of the isotopes, the same book I had used to find the lifetime of Be-10. It was also necessary that the radioactive atom produced by neutron absorption have a long enough half-life so that it could be detected after the sample was removed from the reactor, but not too long, or few of the atoms would decay during the counting period. One can detect only those atoms that decay, emitting a high-energy particle of light called a "gamma ray." The gamma ray had to be unique, and not have the same energy as gamma rays emitted by other kinds of atoms. There would be only a tiny amount of the micrometeorite material in the rock, and only a tiny part of that would be a siderophile element. To a nonexpert the problem would have looked hopeless, but Luie believed his bag of tricks was sufficient to crack the problem; he would find the isotope with just the right combination of properties.

The siderophile element iridium had been named after Iris, the goddess of the rainbow, because of the many colors in the chemical compounds it formed. It is an element that most people have never even heard of, although it is well known to jewelers. Platinum jewelry contains about 10% iridium, added to make the platinum harder and more durable. Luie thought iridium, not gold, platinum, or osmium, would do the job, but he had never done neutron activation analysis. The next step was to find an expert, someone who could alert him to the potential pitfalls. Luie already had an expert in mind.

Frank Asaro was well known among experts in neutron activation analysis, but he had gained public fame with his analysis of "Drake's plate." This was a crude brass plate that had been found in California, near the coast, and it had become a treasure of the university's Bancroft Library. Inscribed on the plate were these words:

BEE IT KNOWNE VNTO ALL MEN BY THESE PRESENTS
   IVNE 17 1579
BY THE GRACE OF GOD AND IN THE NAME OF HERR
MAIESTY QVEEN ELIZABETH OF ENGLAND AND HERR
SVCCESSORS FOREVER I TAKE POSSESSION OF THIS
KINDGOME WHOSE KING AND PEOPLE FREELY RESIGNE
THEIR RIGHT AND TITLE IN THE WHOLE LAND VNTO HERR
MAIESTIES KEEPEING NOW NAMED BY ME AN TO BEE
KNOWNE VNTO ALL MEN AS NOVA ALBION.
                                  FRANCIS DRAKE

In his book *World Encompassed by Sir Francis Drake*, Drake mentioned that he had left a plate on the coast. And after nearly 400 years, it had been found. Or had it? There were skeptics who claimed the printing wasn't authentic, and that Drake never would have left such a crude object.

That year, 1977, James Hart, director of the Bancroft Library, had decided to contribute to the quadricentennial celebrations of Drake's voyage by commissioning a new series of studies of the authenticity of the plate. He asked Glenn Seaborg, a Nobel laureate in chemistry and one-time director of the Atomic Energy Commission, whether anyone at the Lawrence Berkeley Laboratory could drill small samples from Drake's plate without damaging it, in order to send them for analysis to the Research Laboratory for Archaeology and the History of Art at Oxford University, in England. Seaborg suggested that Frank Asaro might like to do the work because of his interest in archeology. Frank, in turn, suggested that nuclear chemist Helen Michel head the sampling operation. Frank and Helen realized that they could test the authenticity of the plate themselves by using neutron activation analysis, and they received permission from Hart to keep half of the samples for their own studies. After nearly eighteen months of study, they concluded that the composition of the metals in the brass was unlike that of any brass used at the time of Drake. The very low amounts of nickel, iron, lead, silver, gold, arsenic, and antimony in the plate could not possibly have been attained in the sixteenth century, at the time of Drake's voyage, because necessary procedures for purifying the metals had not yet been developed. The technology to produce that particular alloy of brass had not existed until the late nineteenth or twentieth century. Several other studies on the plate led to the same conclusion.

The plate was a fake. No one has ever found out who perpetrated the hoax, but neither has anyone countered the compelling evidence that Frank and Helen found. They, in a sense, had taken the fingerprint of the plate and found it guilty. The University of California lost a treasure. The plate is still displayed in the Bancroft Library, but no longer as a famous historical artifact, only as a famous forgery. Universities should never be proud of their museum pieces anyway. Far better that they be proud of their scientists and scholars.

Luie often said that the Lawrence Berkeley Laboratory was unique in the world. Where else, after you decided you needed a new technique, would you just happen to find the world's expert literally next door? Frank and Helen worked in Building 70, right across from Building 50, where Luie worked.

So Luie and Walter told Frank about the clay layer and showed him the calculations. Frank was fascinated. He immediately agreed that iridium was the element of choice, but he was concerned about a possible conflict of interest. He was already collaborating with Andre Sarnov-Wojcicki, of the U.S. Geological Survey, on a project to measure iridium in soils as an indicator of meteor showers. Luie was surprised to learn that somebody else was planning to use iridium to measure micrometeorites. He had invented the idea independently, and then gone to the same expert (Frank) as had Sarnov-Wojcicki. After discussing the project with Frank and Walt, Sarnov-Wojcicki agreed that there was no conflict of interest, and Frank joined the Alvarez team.

I was left out partly because I had nothing more to contribute, but also because I didn't want to get involved. I couldn't afford to waste time on another null experiment. In addition to the cosmology and cyclotron work, I had begun a project designed to eliminate atmospheric distortion in astronomy by flexing a telescope mirror in real time to keep up with the flickering of the Earth's atmosphere. In retrospect, it doesn't sound as exciting as solving the riddle of the dinosaurs' demise. But it was my own project, and I was convinced I could make it work.

The probability of success in Luie's dinosaur project seemed small. Even if he could determine the sedimentation rate, what then? It's interesting how quickly your attitude toward a problem can change when you're no longer involved.

Life itself was the biggest distraction. Rosemary and I had decided to step into the real unknown, to plan an adventure far more daring than anything we had yet tried. After ten years of marriage, we were ready to expand our family.

Meanwhile, Luie and Walt and Frank began looking for iridium.

6. Supernova

IN EARLY 1978, Luie was still waiting for Frank Asaro to do something about his iridium idea. Seven months had passed with no experimental results. Walt had given Frank the clay samples from Gubbio in 1977. Unfortunately, part of the equipment that Frank used in the neutron activation work had become defective and had had to be replaced. This took several months, and then a backlog of 300 samples for other projects had to be measured before the new work could begin. Eventually, the clay samples were irradiated for several hours with neutrons from the small research reactor near the Berkeley campus. The radioactively hot samples then had to sit quietly for a few weeks, while the intense short-lived radioactivity from other elements died off, before the clear (but weak) gamma-ray signal from the activated iridium could be measured.

It is hard to remain excited about any project that takes so long, and Luie's thoughts were mostly on other things. He was thinking of starting a new company to develop further his ideas on stabilized optics. He wanted to maintain complete control of the enterprise, to design, build, and sell high-power stabilized-image binoculars that were light and inexpensive, binoculars that could be used at football games. It would be nice to make a few million dollars, he thought. With his long list of important discoveries and inventions, he had had the luxury of spending his last few years on the long shots. But the trouble with long shots is that they usually fail, and then you have nothing to show. His ideas were brilliant, but people don't like talk about brilliant ideas. They want *results*.

On June 21, 1978, Frank finally had iridium measurements he felt he could trust. Luie told me about them that same day. There was iridium in the clay layer, as he had predicted, at about one part per billion. In the limestone above and below the clay layer, there was no detectable iridium, meaning a level at least 10 times smaller. And there was other good news. Walt had decided not to go back to Lamont; he would accept the appointment at Berkeley. Luie was elated. Walt's decision to stay, not the iridium discovery, seemed to be the big news in Luie's life that day.

On the next day, I flew off to Dallas to receive the Texas Instruments Foundation Founder's Prize, awarded for work I had done in astrophysics, optics, and for the invention of accelerator mass spectrometry (the technique that had failed to have enough sensitivity to examine Walt's clay layer). Luie flew down later that day for the award ceremony. At the party I met the officers and scientists of Texas Instruments, developers of the first silicon transistors and chips, as well as members of the Foundation. But I found myself surrounded mostly by their wives as I vividly described my participation in the natural childbirth of my daughter.

When Luie and I walked through the streets of Dallas the next evening, he said to me that he thought our roles had reversed; he felt he was now learning more from me than I was from him. Luie prided himself on his scientific honesty, and he took great pleasure in my success, but I felt a little sad to see the apparent decline of such a great man. His career had moved so fast, and then come to a virtual standstill.

When Luie returned to Berkeley he learned that the measurements of the iridium layer had not solved Walt's sedimentation-rate problem after all. In fact, there was a new paradox. Frank had found too much iridium, much too much.

Walt had guessed that the production of limestone had simply stopped during the catastrophe at the end of the Cretaceous period. Even when no limestone is forming, silt from rivers continues to deposit clay in the oceans. Walt had expected that the rate of clay from this source would remain constant even when the limestone deposition stopped. Since the iridium was also laid down at a constant rate, he had predicted that the ratio of iridium to clay would remain unchanged. The measurements showed that the iridium level at the Cretaceous boundary increased by

300, but the clay concentration increased by a factor of only 10 (from 10% clay entrapped in limestone to 100% clay). There was 30 times too much iridium.

There was no obvious explanation. Was it worth trying to figure it out? It could be just too hard a puzzle to solve. Even Einstein once bit off more than he could chew. He had wasted much of his career trying to create a unified field theory, but it had turned out to be too difficult; not enough was known at the time. The iridium excess was either trivial or too difficult to explain. There must be lots of sources of spurious iridium, I thought. There are veins of gold in rock; why not veins of iridium? ("This is sedimentary, not metamorphic rock," Walt might have explained had I asked him this question directly.) Maybe a volcano erupted and deposited iridium from the Earth's core. ("Volcano emissions originate from just beneath the Earth's crust, and numerous measurements have shown that their lava is not enriched in iridium," Walt might have said.) A factor of 30 in an obscure element like iridium seemed unlikely to mean anything important.

Luie felt differently. He thought the iridium excess was a new clue, unlike any that had been found before. He sensed a great discovery. Like a shark smelling blood, he sensed something worthy of attack, and he went after it with all his vigor. It was the clue that could unravel the mystery of the dinosaurs. He would find the explanation, somehow.

While Luie worked as a theorist, looking for an explanation for the iridium excess, Frank Asaro continued to make more measurements, to look for more clues. Frank asked his colleague Helen Michel to join the team.

I returned from a trip in August and found that Luie was still applying his skills and energy to the iridium mystery. He felt he was hot on the trail, but he was not yet satisfied that he had found an answer. It still seemed to me that he was wasting his time, just so he could work with his son. When he wasn't in his office, he could be found in the library, often reading a thirty-year-old article. He sometimes stopped by my office with some wild new explanation, and I listened courteously and tried to find flaws. I never could. Luie had always anticipated my objections and had answers ready. He seemed to take delight in the fact that nobody could kill his latest ideas but himself. I could not compete with Luie Alvarez, the Young Turk.

By the end of the summer, Luie had concluded that there was only one acceptable explanation for the iridium anomaly. All other origins could be ruled out, either because they were inconsistent with some verified measurement or because they were internally inconsistent. The iridium had come from space, he decided, and the only possible source was a supernova, an exploding star.

Supernovas are rare events in the heavens. In a typical galaxy of a hundred billion stars, there is one supernova every 25 to 100 years. The last known supernova in our own Milky Way galaxy was 300 years ago, but it is thought that there have been many more that have been obscured by bands of dust near the sun. (In early 1987 a particularly bright supernova in the nearby Large Magellanic Cloud would make headlines across the world.) Infrequent though they are, they are worth waiting for. In just a few weeks the exploding star gets a hundred billion times brighter, until it is as bright as all the other stars in the galaxy put together. It is possibly the greatest release of energy in the modern Universe. Supernovas create heavy elements, such as iron, gold, and iridium, and disperse them into space. Although supernovas are rare, the galaxy is old, and there is virtually no region that has not been hit by such an explosion. Most astronomers believe that the sun was created from the remains of several supernovas. The fact that we have iron in our blood means that we, the atoms we are made of, were once inside a star that exploded. You and I are supernova remnants.

A nearby supernova had killed the dinosaurs. It was Mal Ruderman's old theory, the one that Walt had told me about two years earlier. Luie hadn't foreseen that the iridium would be a verification of this theory, but now it seemed to fit right in. Iridium is produced and dispersed in supernova explosions. The energy from such an explosion, if it occurred in a nearby star, could have swept away the atmosphere of the Earth, or simply created a blast of heat sufficient to kill life. Nearby supernovas aren't common events. Usually you would have to wait several billion years for one that could affect the Earth. But rare events do happen. It was reasonable to think that the destruction of the dinosaurs, a unique kind of event, could have been triggered by a rare nearby supernova explosion, particularly since it could account for the iridium anomaly.

(Luie had no idea, at the time, that a few years later we would believe that catastrophes such as the one that killed the dinosaurs are not rare, but in fact happen on a regular periodic timetable.)

Another scientist might have published the new iridium data, explained how it was consistent with the Ruderman supernova theory, and sat back content that he had solved the riddle of the dinosaurs. But in Luie's mind a theory was no good unless it could predict something new. What could he look for that would test the supernova hypothesis? It didn't take him long to figure it out: plutonium.

Plutonium is called an "artificial" element because it is almost absent in the Earth's crust; most of the world's plutonium is manufactured from uranium in nuclear reactors. The main isotope of plutonium, Pu-239, decays with a half-life of only 24,000 years, so any of this isotope that was present on the Earth at creation has long since decayed away. But Luie knew that there was another isotope of plutonium, Pu-244, which has a much longer half-life, about 80 million years. The supernova explosion should have created new Pu-244, and injected it, along with the iridium, into the Earth's atmosphere. Only 65 million years have passed since then, and most of the Pu-244 would still be left. If it could be found in the clay layer, it would prove the theory. Not just verify it, not just strengthen it, but prove it. There is no other conceivable source of Pu-244. Its presence would establish the supernova explanation beyond all reasonable doubt.

How could Pu-244 be found? Its half-life of 80 million years was so long that very few atoms decay per hour, so it could not be detected from its own radioactivity. Luie asked Frank and Helen if they could find it using neutron activation analysis. They said they could, but it would be very difficult. When Pu-244 absorbs a neutron, it becomes Pu-245. This has a short half-life, only 10.5 hours, so half the atoms would emit radiation in that period. They could detect the presence of Pu-244 by exposing the sample to neutrons in a reactor, and then looking for the characteristic radiation from Pu-245. But the half-life of Pu-245 was so short that most of the characteristic radiation from this isotope would quickly disappear. The neutrons from the reactor would make virtually every other element in the sample radioactive, not just the minute amount of Pu-244. Since the half-life of Pu-245 is so short, they could not wait for the other radioactivities to die away, as they had in their iridium measurements. They would have to make the measurement with the radioactively hot sample. Chemical procedures would be necessary to concentrate the plutonium as much as possible. Helen's skills as a radiochemist, someone who can safely do chemistry with very radioactive substances, would be particularly critical.

Frank and Helen were not exaggerating when they called the measurement "very difficult." When Frank later described the procedures to me, I found them almost incomprehensible. Their first task was to concentrate the plutonium as much as possible. They spent two weeks doing this to 25 grams of material that Walt had collected near Gubbio. They treated the clay with hydrochloric acid, to dissolve away whatever limestone was mixed in. A small amount of another isotope of plutonium, Pu-238, was added ("spiked") to the sample. Since the chemistry of this isotope is identical to that of the desired isotope, Pu-244, they could use it as a tracer to calibrate how well their chemical concentration was working. The spiked clay was heated with hydrofluoric and nitric acids to destroy the silicate structure. The silicon came off as silicon tetrafluoride gas, while many other elements (including the plutonium) were left behind as insoluble fluorides. The residue was washed with dilute acid and then fused with solid potassium hydroxide to break up the fluoride precipitates. Next they cracked the fused mixture into pieces and treated these with a solution of lye to dissolve away the alumina, a major constituent of clay. The insoluble hydroxides left behind were dissolved in phosphoric acid with a little nitric acid and a reducing agent. This process left the plutonium in a triply ionized state (+3), making it behave chemically like a rare earth.

The solution was heated and stirred, and bismuth nitrate slowly added. Bismuth phosphate precipitated in solid form, carrying with it the plutonium and other rare earths (the lanthanides). It was a procedure that had been devised by Stanley Thompson during the Manhattan Project for the purification of plutonium, and had been of immense use in the discovery and study of many new elements.

The bismuth phosphate precipitate was dissolved in hydrochloric acid, and a few drops of lanthanum chloride solution added. Lanthanum fluoride precipitated as a solid, carrying with it the plutonium (which was behaving chemically like a lanthanide). The solid was washed to remove the bismuth phosphate. Lanthanum fluoride was removed by dissolving it in boric acid, and lanthanum hydroxide was precipitated by adding a base, which removed the boric acid. The precipitate was dissolved in nitric acid, which also changed the ionization state of the plutonium from +3 to +4. In this state the plutonium no longer behaved like a lanthanide, and it could be separated by running the liquid through a tube known as an anion exchange column. The plutonium became stuck in the column, and was later removed by running through the column a moderate concentration of hydrochloric acid containing hydroxylamine as a reducing agent. This reduced the plutonium back to the +3 ionization state, and allowed it to be washed out. By measuring the remaining amount of the

Pu-238 that had been spiked into the sample, Frank and Helen were able to determine that about half of the original plutonium had been recovered from the original 25 grams of rock.

On March 6, 1979, after this initial purification process had been completed, the plutonium fraction from the clay was irradiated for eight hours with neutrons in the small research reactor near campus, to convert as much as possible of the Pu-244 into Pu-245. (A longer exposure would not have helped much. After eight hours the Pu-245 was decaying almost as fast as it was produced.) Even though the sample was now very small, it still contained many radioactive contaminants, especially lanthanum. More chemical purification was needed, but now time became critical; half of the signal from Pu-245 was disappearing every 10.5 hours. There could be no stopping until the contaminants in the very radioactive sample were removed. (The amount of plutonium would be too small to be a hazard, but care would be needed to handle the radioactivity of the other elements.) Frank and Helen dissolved the material in hydrochloric acid with a little nitric acid as an oxidizing agent, and ran the solution through an ion exchange column. They worked all night, while Luie, Walt, and Walt's wife, Milly, brought them coffee and snacks. Periodically, Frank and Helen put the material into their gamma-ray detector system and measured the residual background. At 5:00 A.M. there were still detectable amounts of lanthanum and other contaminants. So a new ion exchange column, just like the last one, was run. Finally there was a period of 10 minutes with no counts above background, an indication that their purification method had removed most of the interfering elements. It was time for the radiation counters and the electronics to work. Frank and Helen could sleep.

Six hours later they looked at the signal that had built up in their analyzer. They looked at the peaks in the pulse-height spectrum, and identified the elements in the sample. There were numerous cross-checks to make. If there was a true plutonium signal, then there must also be an americium signal, since Pu-245 decays into americium. Plutonium without americium is fool's plutonium.

I saw Luie and Walt the next day in the hallway of Building 50. Luie said they were writing a paper on the supernova "possibility," even though Frank and Helen's plutonium measurements had not yet been completed. I was horrified. I played senior statesman, taking the role I had expected Luie to play, and told them it would be a big mistake to write a paper now. Wait until you see plutonium, or show that it isn't there, I said. You are so close. Don't abandon your standards now. Luie said that he would seriously think about my advice.

The next day I was talking to some colleagues in my office when Luie came in, surreptitiously slipped me a little note, and left. It read, "Meet me in the hallway." A minute later I was out there. "We found the plutonium!" Luie whispered to me. "Walt and I knew it yesterday, when we talked to you. I misled you when I made you think we hadn't found it yet. That's why we're writing the paper now." Why hadn't he told me yesterday? "Because I wanted to show Walt how important it is to keep this a secret," he said. "Walt knows that if I don't tell something to you, it means I won't tell it to anyone."

Luie filled in some of the details of the new results. There was absolutely no doubt about the presence of plutonium-244. Several distinctive gamma rays had been detected, with the right energies and ratios of intensities. They had even found the americium, the ashes of decayed plutonium. Frank and Helen had performed magic, and their results were indisputable. There was Pu-244 in the rock. A supernova had killed the dinosaurs.

But why keep it a secret? Why not tell the world? I was one of the very few people who knew, really *knew* what killed the dinosaurs! It was the most exciting fact that I had ever been privy to. Why not broadcast the news? Luie had several answers. Frank wanted to check everything, to be doubly sure; there was still the outside chance of a mistake. Their discovery would get the attention of so many people in the scientific world that a mistake would be dreadful. Besides, there might be other things to exploit, other consequences. It was very rare to have such exclusive knowledge. Maybe there was some important consequence of the supernova theory that we could still find, now that we knew we were right. There was no rush to publish, not yet. A lot of people knew about the iridium by now, but nobody except us knew about the plutonium.

I was tremendously excited. I knew the rules of the game, and I knew that it would be okay to tell my wife, Rosemary, about the plutonium and how it proved the supernova theory. I knew that wives don't count when Luie says "nobody"; except for classified defense secrets, Luie had never kept anything from his wife. Unfortunately, Rosemary wouldn't be home until 5:30. Then I had an inspiration! It was March 8, my sister Virginia's birthday. What better present than to let her be one of the first dozen people in the world to learn what had killed the dinosaurs? Are sisters included in the okay list? I was sure I could trust her. She wouldn't go off and publish the theory herself. Of course, I would have to make her promise not to tell anybody. I went to visit her and told her what had killed the dinosaurs. She seemed genuinely excited.

What about Mom and Dad? They had recently moved to Berkeley, to be closer to us and to Virginia, but mostly to be closer to their first grandchild, Betsy. We had found them an apartment two blocks from us. How could I *not* let them share in this news? They had been hearing the name Luie Alvarez from me for thirteen years. I was sure they wouldn't tell anyone if I asked them not to. I stopped by their apartment to tell them. They seemed to be excited by the news.

Rosemary finally came home, and I told her. She was more skeptical than excited. "If they are so sure, why are they keeping it a secret?" she asked. I tried to explain, but under her perceptive questioning my reasons didn't really add up, even to me. I finally retreated to authority: "Because Luie wants it kept secret, for now." Betsy was only nine months old. Infant daughters were also okay to tell, I assumed, and I told her next. She didn't seem to understand.

Luie's elation could hardly be contained, even though it would take Frank and Helen a few weeks to check the result, to repeat the entire measurement. Luie called Frank Press, an old friend, and at that time President Jimmy Carter's science adviser. He told Press of the discovery and suggested that Press invite him to give a talk to announce the discovery at the National Academy of Sciences meeting coming up in November. Press immediately agreed to arrange the talk. Luie knew that the world would want to hear more details; he thought a good second forum would be the annual meeting of the American Association for the Advancement of Science, planned for January in San Francisco. He was on the local organizing committee for this meeting, along with David Saxon, then president of the University of California, and Sherwood Washburn, a distinguished anthropologist. He shared his secret with them, and they were delighted to accept his offer to give the talk that would obviously be the highlight of the meeting. Walt was planning a trip to Denmark to attend a conference that would discuss the extinctions, and Luie decided to go along. He would enjoy helping spread the news worldwide.

Frank Asaro is an extremely cautious scientist. The evidence was very strong that he and Helen Michel had actually seen plutonium-244, but he wasn't satisfied. The gamma-ray lines from americium would have convinced every skeptic that the plutonium was there; however, there were other checks Frank could make, and he intended to make them. Six days after their original measurements, he and Helen repurified the sample in which the discovery had been made, and exposed it again to the reactor. When they then measured the gamma spectrum, they could again see the lines characteristic of Pu-245 and Am-245, just what was expected from neutron-activated Pu-244, and confirming that their original measurements were correct. Once again there was absolutely no doubt. The sample contained plutonium.

The level of plutonium was extremely low. Trivial as that fact seemed, it bothered Frank. He didn't want to stop until he understood every aspect of the measurements. Could they be sure that the plutonium had come from the clay? Where else could it have come from? He knew that plutonium had been used at the Lawrence Berkeley Laboratory before. Had enough care always been taken to avoid all possible contamination? Lab safety precautions were tight and had always proved effective. The amount needed to produce the signal in the clay sample, however, was far below the danger level, even by the standards of the most ardent anti-nuclear activists.

One recollection particularly bothered Frank. During the measurements his supply of hydrofluoric acid had run out, and he had borrowed some from an upstairs laboratory. He checked the history of that bottle of acid, and found that it had been used on a special table with its own exhaust hood, used for potentially dangerous substances. Had Pu-244 ever been used on that table? Unfortunately, the answer turned out to be yes. Could the hydrofluoric acid have picked up a minute, almost unmeasurable amount of Pu-244 from the hood, and then transferred some to the clay sample? Probably not.

Probably not. That was not good enough for such an important discovery. Frank and Helen decided they had to repeat the *entire* experiment, this time using fresh hydrofluoric acid. They got ready to repeat the measurements on two independent samples of the boundary clay, each twice as large as the first. All of the tedious procedures would have to be repeated. Besides new clay samples, there would be new beakers, new instruments. Luie recognized the value of Frank's care, even though he never would have exercised so much himself.

Frank and Helen spent two weeks concentrating and purifying the new samples prior to irradiation. After they were exposed in the reactor, the same chemical procedures were used as before, ending with two anion exchange columns. Their yields this time were more than twice as good as before, so they expected a very strong plutonium signal. They put one sample into their system and left the detectors to continue running for seven hours. They then put the second sample in for nearly twenty-six hours. To their dismay, there were no gamma rays at the Pu-244 energies. Frank looked for the americium decay gammas. They were absent, too. There was no Pu-244.

They made one more check. If the Pu-244 was due to minute amounts picked up from the hydrofluoric acid, there would be other isotopes of plutonium present. Maynard Michel, Helen's husband and a Lawrence Berkeley Laboratory nuclear chemist, measured the relative amounts of the plutonium isotopes with a mass spectrometer. His measurements of isotope ratios showed clearly that the Pu-244 had come from the upstairs laboratory.

The previous result was wrong. There was no plutonium. A supernova had not killed the dinosaurs.

These results were later published in a paper titled "Negative Results of Tests for the Supernova Hypothesis." In this paper they described how the measurements showed that there was no plutonium-244 present, and how this result ruled out the supernova theory. In order to demonstrate what a verification of the supernova theory would have shown, they gave a plot of the data from the first measurement. The sample had been "spiked" with plutonium, they said. Nowhere in the paper did they admit that the spiking had been accidental. And nowhere did they mention the excitement they had experienced over this initial but incorrect result.

The very worst thing that an experimental physicist can do is announce an important result and then have it proven wrong. Luie told me, "Rich, there are several people in the world to whom I give credit for having saved my life. In that list I now include Frank and Helen."

7. Asteroid

NO PLUTONIUM meant no supernova. One clear fact, solidly established, is enough to disprove a theory. There was no way to salvage it. The supernova theory was dead.

Yet progress had been made. Sometimes, when you find that you have placed a piece of puzzle in the wrong place, the discovery allows you to fit whole new sections of the puzzle together. Many scientists would have been proud of disproving a theory by a scientist like Ruderman, one of the most respected theorists in the world, but not Luie. For him, the only worthwhile goal was *the* solution. The iridium was still there, daring him to explain it. This was not a failure, but an opportunity. In his book, *The Structure of Scientific Revolutions*, writer Thomas Kuhn had used the word *paradigm* in his analysis of scientific revolutions to describe the conceptual framework into which we try to place new scientific discoveries. Those that *don't* fit are often the most exciting, for they imply that we must change the paradigm itself. Copernicus had changed our understanding of the role of man in the Universe when he showed that it did not revolve around the Earth. Einstein had changed our paradigms of space, time, and energy with his theory of relativity. Heisenberg had altered our concept of reality with his uncertainty principle. What new paradigm was needed to explain the iridium? What was wrong with our thinking?

Even more than before, the shark in Luie smelled blood. He attacked the problem with almost unbelievable vigor. He told everyone who was interested that he was trying to solve the most exciting problem of his life, but I don't think many people took him seriously, not even Frank, Helen, or Walt. I know *I* didn't believe him, despite the brief flurry of excitement over the supernova explanation. When he got his teeth into a problem like this, Luie didn't let go until he had totally ripped it to shreds.

Meanwhile, I began to have all those wonderful distractions that come with a modicum of national recognition. Frank Press asked me to be on a special committee to investigate a report that South Africa had tested a nuclear weapon. (We were able to show that they had not made such a test.) Robert Frosch, the head of NASA, asked me to join a special "innovation panel" to look into the long-term future of his agency and suggest projects for the 1990s and beyond. My radioisotope technique was progressing rapidly, and more than a dozen laboratories around the world were using it. I took a trip on a Lear Jet owned by Mark Hungerford, a local entrepreneur, to see the eclipse of the sun in Montana. Robert Budnitz, an old friend from Berkeley now in charge of research for the Nuclear Regulatory Commission in Washington, called to tell me the inside story of the Three Mile Island nuclear-reactor accident just days after it happened. My recently acquired tenure position in the Physics Department at Berkeley meant that I was financially secure. It was a good life.

In contrast, Luie was sixty-eight years old and well into his retirement, but he was attacking the iridium problem as if he were twenty. He seemed to enjoy the stress he was putting into his life.

He still felt that the iridium must be extraterrestrial. If not a supernova, what else could deliver it? One possibility was an asteroid, a chunk of rock several miles across. Christopher McKee, a theoretical astronomer at Berkeley, had suggested to Luie that if an asteroid hit in the ocean, a huge tsunami, a tidal wave, would sweep over the continents and kill the life there. Luie realized that the tsunami would move around the continents and pass several times around the world. It was a good theory, but when he looked at it more closely he decided that the waves would not reach the middle of the continents. Dinosaurs in Montana and Siberia would survive. It was also hard to see how the tsunami could be so devastating to ocean life. He concluded that the theory didn't work. McKee agreed. Luie looked for new ideas.

Perhaps instead of thinking about the origin of the iridium, he conjectured, it would be productive to think about what could have killed the dinosaurs. The iridium itself could have had no effect; it was only a minor element, useful as a tracer. The most common element in the Universe is hydrogen. Whatever brought in the iridium might also have brought in a vast amount of hydrogen. If a large cloud of this gas had hit the Earth, it could have combined with the oxygen in the atmosphere to make water vapor. Sunlight, acting as a catalyst, would have made the reaction take place rapidly. With the oxygen all used up, the dinosaurs would have been asphyxiated. How could this theory be tested? Was there any evidence that the oxygen of the Earth had or had not gone away? Luie concluded that there wasn't.

Maybe the hydrogen had come from our own sun. Luie knew that there were sometimes giant solar flares, knots of hydrogen plasma tangled in magnetic fields, that erupted from the surface of the sun. Such flares had posed a constant threat to our astronauts, and during their time in space a special solar-flare watch had kept an eye on the sun. The astronauts were prepared for an emergency landing in the event of a giant flare. How big a flare can you get if you wait 65 million years? Could we get one so huge that the hydrogen would destroy the Earth's oxygen?

Luie told me this idea, and I decided to help by looking into it on my own. Under "Solar Flares" in George Abell's introductory book on astronomy, *Exploration of the Universe*, I found a list of novae, stars that undergo a relatively small explosion similar to a giant flare. I realized that the list contained stars very similar to the sun. So maybe the sun had been a nova. The catastrophe might not have been due to a supernova of a nearby star, as Luie had once thought, but due to an ordinary nova of our own sun. The lack of plutonium didn't rule out an ordinary nova. I showed the table to Luie, who immediately became interested. What were the consequences? Were there any data that could rule out this possibility?

Luie studied the nova possibility with great care. He learned that all the stars in the nova list had companion stars, orbiting very close to their surfaces. Novas come about because the companion star slowly leaks material to the primary star, and when enough excess material accumulates it undergoes a sudden thermonuclear detonation, resulting in the nova. The sun has no such companion, so it could not become a nova. Another week, another theory dead.

Suppose the Earth happened to pass through a giant molecular cloud. These large clouds, floating among the stars, had recently been discovered by radio astronomers. Perhaps the dust in such a cloud blocked the sunlight from reaching the Earth. The great astronomer Fred Hoyle has written a science fiction story in which a dust cloud blocked sunlight and caused the Earth to chill below freezing. But that was science fiction, and Luie soon calculated that the dust in the known giant molecular clouds wasn't dense enough to serve. Could such a cloud, instead, bring in enough hydrogen to use up the Earth's oxygen? It didn't matter, Luie realized, because he found another way to rule out this theory. At the rate that the Earth is moving through the Milky Way galaxy, a mere 20 miles per second, it would take far too long to pass through the clouds to give the observed iridium layer.

Luie was going strong, and keeping ahead of everybody. The short-lived nova theory was the only contribution I made during this period. I couldn't compete with his broad experience, wide knowledge, and concentration. He believed he was bound to find the answer, if only he could keep on coming up with one good new idea every week. He just had to keep at it. He had to be tougher on himself than on anyone else. Keep thinking.

Jupiter has a lot of hydrogen. What could bring that hydrogen to the Earth? Jupiter is near the asteroid belt. Could an asteroid strike the belt, releasing hydrogen? How could the hydrogen get to the Earth? Luie could not get this idea to work.

He had not thought hard about asteroids since he and McKee had abandoned the tsunami idea. He had spent most of his time trying to work out a model that would bring excess hydrogen into the atmosphere. He had gotten nowhere, except to eliminate a lot of clever ideas. But he finally returned to the possibility of an asteroid. Even a small telescope shows that the moon is covered with craters. The Apollo missions had proven that most of these were from the impacts of asteroids or comets. Suppose an asteroid came close to the Earth and passed horizontally through the Earth's atmosphere. Maybe it would break up and create a cloud of dust that could block the sun, just as in Hoyle's science fiction story. If it became trapped in a shrinking orbit around the Earth, the atmospheric drag might do it. This idea lasted a few days, and Luie told several people about it with his characteristic enthusiasm. But then he calculated that the atmospheric drag would be no more than the gravitational force of the Earth. It would be too weak to break up an asteroid.

Suppose the asteroid hit the Earth? The tsunami wouldn't get inland, but what else might happen? The impact would make a crater, like Meteor Crater in Arizona. Meteor Crater is about a mile across and 30,000 years old. In the 4.5-billion-year history of the Earth, a much larger object could have hit, creating a much larger crater. How big could it have been? Vertebrates had been on Earth for only a couple of hundred million years. How big an asteroid might have hit the Earth during that time? Luie had read a recent *Scientific American* article by G. W. Wetherill that talked about "Apollo objects," asteroids whose paths cross that of the Earth. He reread it, and then went to the library to look up a more detailed article in the *Annual Review of Astronomy and Astrophysics*. A quick extrapolation from the Apollo objects listed in this article indicated that the largest one that might hit the Earth in a hundred-million-year period would be about 5 kilometers in diameter. Another possibility would be a comet, with a core of comparable size. It was somewhat more likely that the Earth would be hit by an asteroid than by a comet.

The asteroid would crash into the Earth with a velocity given by the vector sum of its velocity with that of the Earth. The Earth moves around the sun at 30 kilometers per second, over 10 times faster than the fastest bullets and artillery shells. An impact at that speed would carry a great deal of energy, but not much compared with the kinetic energy of the Earth. An asteroid diameter of 5 kilometers is 2,000 times smaller than that of the Earth; its momentum is proportional to its mass, which is proportional to the cube of the diameter. Thus the asteroid would have a mass 10 billion times smaller than that of the Earth. The impact could not alter the orbit of the Earth by more than 93 million miles (the distance from the Earth to the sun) divided by 10 billion—less than 50 feet.

Kinetic energy is proportional to the square of its velocity, so each gram of the asteroid would carry 100 times more energy than the explosive (comparable to TNT) that drives a bullet. (The same principle is used for the U.S. antisatellite weapon, which carries no explosive. Direct impact of the mass at these high velocities releases more energy than the explosion of a comparable mass of TNT.) The asteroid would have a mass of about a million megatons; multiplying this by the factor of 100 (for the high velocity) shows that the impact would release the energy of 100 million megatons of TNT. That is 10,000 times greater than the *entire* U.S. and Soviet nuclear arsenals.

What happens when that much energy is suddenly released at one place on the surface of the Earth? There was no recorded event in human history with comparable destructive power. Luie looked up data from the largest nuclear explosions, and tried to extrapolate. He checked his numbers by looking at the estimates astrophysicists had made for impact craters on the Earth and the moon. The material of the asteroid would be heated by the impact to over a million degrees Celsius. It would vaporize surrounding rock, melt even more, and eject a huge amount of material into the atmosphere. The impact would make a crater 100 miles across. The material thrown into the atmosphere and above would eventually cool and turn into dust. If it took long enough to settle down through the atmosphere, it would be spread out over the entire Earth by high-altitude winds. There would be enough material to block the sunlight. That must be what did it! The sunlight went away. Mount St. Helens had recently erupted in Oregon, darkening skies over nearby towns. For how long would the darkness from an impact last?

Luie knew about Krakatoa, the volcano in the South Pacific that had exploded in 1883 and had thrust dust and rock over 30 miles into the atmosphere. The dust had spread worldwide, creating beautiful reddened sunsets at distant locations for years. The dust from the asteroid impact would undoubtedly settle in a comparable time: several years. That would be long enough to kill most life. In fact, it would be something of a miracle if anything made it through.

Luie's estimate of several years for the dust to settle, based on the red sunsets from Krakatoa, was one of his few errors. It wasn't really his fault, since he took the estimate straight out of the Royal Society's century-old summary of everything that had been learned about the Krakatoa explosion. It wasn't as serious a mistake as the one I had made in using the wrong lifetime of Be-10, but it later came back to haunt him. Dust from Krakatoa had actually settled out in a few months, although a tiny fraction had remained in the atmosphere for several years, creating the sunsets. Later, paleontologists used Luie's incorrect "several years" estimate as an excuse to dismiss his model as patently absurd. Nothing would make it through a period of darkness several years long, they argued. Or at least virtually nothing; not the 20% to 50% of the species that actually survived. Luie's team had expertise in chemistry, physics, astronomy, and geology, but not in biology.

Would an asteroid impact lay down the right amount of iridium? Luie knew that the fraction of iridium in an asteroid was probably very similar to that in meteorites, about half a part per million. He assumed the density of the asteroid was similar to that of rocky meteorites, i.e., three times the density of water, and calculated the mass of iridium present. It came out right. Spread out over the world, it would give just the level that Frank and Helen had measured. Had it not agreed, Luie would have added the asteroid model to the trash heap of eliminated theories. Then Luie wondered if he could account for the clay layer, not from rivers, as Walt had assumed, but as debris from the asteroid and its crater. Could the clay be the settled dust? He worked out the numbers, and they agreed with this speculation.

Frank and Helen did their best to prove the theory wrong (Walt was in Italy), but without success. It seemed to hold up better than the previous models. After two weeks with no change in the model, I began to wonder if Luie had really found the solution.

Luie pulled out the draft of the iridium discovery paper and began modifying it for the new theory. As he reworked the numbers, he decided that he had four independent ways to find the diameter of the impacting asteroid, and the four agreed. First, by assuming the iridium in the clay layer was world-wide, he could calculate that 50,000 tons of iridium had been laid down on the Earth. Correcting for the amount that was not ejected into the atmosphere, he estimated that the asteroid must have contained 5 times this much iridium: 250,000 tons. The mass of the asteroid should be about 2 million times larger, giving it a mass of 500 billion tons. This would make an asteroid 8 kilometers across, about 5 miles. Second, he knew that about 100 million years passed between large mass extinctions. He looked at the astronomical data for the largest object that was expected to hit the Earth with this frequency, and once again the answer came out to be about 5 miles. (His theory was becoming more revolutionary as he realized he might be accounting for all the mass extinctions.) Third, he knew that the clay layer came from the crater. Enough rock for the layer meant the crater must have been about 100 miles across. Looking up the best estimates for impact craters, he found that this would take an object between 5 and 10 miles across. Finally, he asked how large an object had to hit in order to block out sunlight, worldwide, for an extended period. Once again he obtained the same answer.

Luie's theory was pregnant with predictions. There must be evidence somewhere on the Earth of the crater, 100 miles across and 65 million years old. Unfortunately, Walt pointed out on the telephone from Italy, most of the area of the Earth is ocean. Geologists had never been able to identify impact craters on the ocean floor, largely because of their inaccessibility. (The first impact crater on the ocean floor was identified eight years later, in 1987.) In fact, more than 20% of the ocean floor that existed at the time of the catastrophe had subsequently been subducted, buried under another land mass by continental drift. So we might never find the crater, even as ocean-floor measurements became feasible.

But there were other, equally important, predictions of his new theory. The iridium enhancement must be worldwide. It must be possible to find it in seafloor cores, and at other sites around the world. The clay in the layer should be chemically similar everywhere, and dissimilar from the clay in the limestone above and below, since it came from a distant crater (with 1% to 10% asteroid material mixed in) and not from local rivers. Perhaps there would be iridium layers at the times of other mass extinctions, since asteroids must have hit the Earth many times. (Luie, Walt, Frank, and Helen would spend much of their time for the next several years slowly and carefully checking these predictions.)

Luie gave me an early draft of the discovery paper, and he asked me to "bloody it up." He wanted lots of red ink marks, lots of criticism. This was going to be a classic paper, and he wanted it to be perfect, no mistakes. He needed friends who would give him a hard time, and I was pleased to be one, but there was little to improve. I sensed that my skepticism was fading, and being replaced by jealousy. I had been close to Luie during most of the last decade and had had many opportunities to contribute, but I had not taken advantage of these opportunities. Now it was too late. Had I made slightly more effort, I might have come up with some ideas, or have helped with the measurements. I was sure that Luie then would have welcomed my coauthorship of this paper. Here was one of the greatest scientific papers of the century, and I was just a bystander.

I thought of all the "important" things that had distracted me. The real problem was my judgment. I simply had not ever imagined that Luie was attacking a truly important problem. He had solved the riddle of the dinosaurs! Perhaps I could now earn an acknowledgment by reading the paper carefully. I suggested a few minor changes and a shorter title. Luie accepted my suggestions. When Walt returned from Italy, he extensively rewrote the paper, adding to it a lot of geology.

In the laboratory cafeteria, I talked to Judith Goldhaber, a friend who works in the public relations department at Lawrence Berkeley Laboratory, about Luie's discovery. I told her that this was his most important contribution to science, more important than the work with elementary particles that had led to his Nobel Prize. Judy knew I was a fan of Luie, but she was surprised to hear the intensity of my admiration of this latest work. She asked me to tell her more about Luie's theory. "It's not a *theory*," I answered. "It's a *discovery*. Luie's greatest. His picture is going to appear on the cover of *Time* because of this." Six years later, Judy reminded me of my prediction, which had almost come true. But Luie's picture appeared inside the magazine, not on the cover. On the cover was a beautiful green-and-orange dinosaur, looking angrily over his shoulder at a great mushroom cloud of dust rising in the air from the explosion of a newly formed impact crater.

Luie knew I was taking a trip to a meeting in Washington, D.C., and that I would see Mal Ruderman there. He was anxious to learn how the originator of the supernova theory would react to the new work, so he asked me to deliver a copy of the just-finished paper to Mal. On December 1, 1979, I gave Mal the copy. The next day Mal gave me a note to deliver to Luie, and said I could look at it. It said simply: "Dear Luie: You are right and I am wrong. Congratulations! Mal."

8. Skepticism

"**B**UT I'VE just printed several other papers 'explaining' what killed the dinosaurs!" This was the complaint of Philip Abelson, the editor of *Science*, who didn't want to publish yet another theory. He continued, "Out of these N papers, at least N-1 must be wrong." Walt was having dinner with Abelson at a national meeting and was trying to convince him that his and his father's theory wasn't just another speculation. It was based on a new *discovery*, the iridium enrichment in the clay layer, and it was unique in its ability to explain this enrichment. Abelson finally backed down, but insisted that the length of the article be cut in half.

Luie and his team had found the answer by taking a very indirect path. What had begun as a relatively straightforward measurement of sedimentation rates had evolved into a puzzle that required an extraterrestrial impact for its solution. I was certain that someday the story would be told in textbooks, but in the simpler, more logical way that textbook authors thought it *should* have happened. The fantasy history would illustrate the "scientific method." It might read something like this:

Astronomers had measured the orbits of many asteroids, and they had realized that one was likely to collide with the Earth every hundred million years or so. Such collisions in the past must have traumatized all life on Earth, since the dirt thrown into the atmosphere from the crater would have blocked out sunlight for periods of many months. It had long been suspected that such events were responsible for the great catastrophes in the evolution of life, but there was no direct evidence until Luis Alvarez and his collaborators at Berkeley found an enrichment of the element iridium, characteristic of such impacts, at the Cretaceous/Tertiary boundary.

It could have happened that way. Maybe it should have happened that way. But it didn't happen that way.

It is true, however, that scientists *could* have made the deduction in the way my fictional textbook account describes had they been smart enough. Collisions with asteroids and comets are bound to happen. They must have occurred in the history of evolution. And when they occurred, they must have caused the kind of trauma that Luie deduced. The logic was undeniable. It gave us all a great deal more confidence in the impact scenario than we would otherwise have had. Such events *must* have happened, the only real question was: When, exactly, had they occurred? It was a question I liked to pose to skeptics. Since catastrophes from impacts must have occurred, where in the paleontological record were the extinctions they should have caused? If they didn't want to identify the Cretaceous extinctions with impact, then which extinctions would they pick?

On June 6, 1980, the discovery paper, entitled "Extraterrestrial Cause for the Cretaceous-Tertiary Extinction," finally appeared in *Science*, with Luie, Walt, Frank, and Helen as authors. The article was met with skepticism and even derision. A few scientists, besides me, hailed it as a great work. However, many experts dismissed it out of hand. Had it not been for Luie's Nobel Prize, they might have been able to ignore it completely.

Luie and his team had entered an area of science that was not considered respectable by most other scientists. It was a field that had attracted amateurs, pseudoscientists, well-meaning scientists from other fields, and "nuts." It sometimes appeared to the paleontologists that nearly everyone in the world had tried to explain the extinction of the dinosaurs, usually without doing their homework. Paleontologists had learned from hard experience that the only possible response was simply to ignore the plethora of theories that bombarded them.

The Alvarez work had all the characteristics of a nut theory. It had very little paleontological jargon, so it was easily recognizable as written by an outsider. It talked about physics and geology, with only passing mention of fossils. The group had no paleontologist, just a physicist, a geologist, and two chemists. Why can't those guys stick to their own fields? Give a guy a Nobel Prize, and he thinks he's a world expert on everything!

Many other fields have some area that attracts nuts. In the Physics Department at Berkeley we regularly receive letters claiming to show Heisenberg's uncertainty principle or Einstein's theory of relativity to be wrong. The Mathematics Department frequently receives "proofs" of the famous (but unproven) "Fermat's last theorem." The faculty members usually take turns responding to these letters. Sometimes they assign them to their graduate students as exercises: "Find what is wrong with this theory, and write a polite reply."

With no more information than what they saw in the newspapers, many paleontologists could tell immediately that the theory was wrong. The fatal flaw was obvious. An impact would have destroyed the dinosaurs over a period of a few years or less. But paleontologists "knew" that the dinosaurs had disappeared slowly, over a period of several million years. Many species had died well before the putative impact. "Maybe the dinosaurs were smart enough to know an asteroid was coming, and they died of fright!" one paleontologist jeered. The impact theory was "obviously" wrong, to the experts.

Luie, for his part, knew that the paleontologists were wrong. The iridium could be explained only by postulating an extraterrestrial source. He had worked very hard to find an extraterrestrial mechanism, and had found only one that was consistent with all the facts. (As is common in the world of science, soon after his work, alternative theories were published. But, as near as I could tell, they were all theories or variants of theories that he had rejected, and he knew what mistakes the authors had probably made.) There had been an impact of an asteroid or comet; of this Luie was certain. And that meant that the paleontologists' claim of extended extinctions must be wrong. With both sides so sure of themselves, the scene was set for a long period of controversy.

Walt was more sympathetic to the paleontologists than Luie was. Geology had much in common with paleontology. Not only did both sciences make extensive use of fossils, but they both had to handle complex data and complex phenomena. Unlike physics, these fields rarely had simple explanations to account for their observations. There were certain theories that both geologists and paleontologists believed to be true, but that they would have difficulty justifying to a skeptic, because they were founded on a large number of details that made sense only in the context of the full explanation. Walt listened to the paleontologists with a sympathetic ear. Yet he also understood the details of the impact theory, in some ways better than Luie. He had reconciled Luie's calculations with his vast knowledge of geology. There must be some way to understand the objections of the paleontologists, and to reconcile them with the impact theory.

Luie was proud to be an iconoclast, yet he was particularly pained by the way he was branded a "nut." In a newspaper interview, William Clemens, a paleontology professor at Berkeley, referred to the impact theory as "codswallop." (This fascinating word sent Walt and me to the dictionaries, in a futile attempt to understand the metaphor.) Richard Casanova, editor of *Fossils Quarterly*, wrote a letter to the *New York Times* in which he called the Berkeley work a "nutty theory of pseudo-scientists posing as paleontologists." Luie readily admitted that he was no expert in paleontology, but it didn't matter. To dismiss his theory, you must have an alternative that can explain the iridium. The iridium was striking, dramatic evidence that could not be ignored. Luie was shocked when he attended a meeting at which paleontologists discussed the mass extinctions and discovered that nobody there even mentioned the word *iridium*. He concluded that they must be afraid of it. Several revolutions in physics had come about because of new discoveries that persisted and were difficult to explain. This was the beginning of a revolution in our understanding of the Cretaceous catastrophe, perhaps of evolution. *The iridium could not be ignored*.

A friend of mine told me the jokes the graduate students were telling in the Berkeley Paleontology Department. "Alvarez is so contaminated with iridium," they said, "that he glows in the dark." In this case I felt that Luie had the last laugh, since the students apparently didn't even know that iridium is not radioactive.

These paleontologists still subscribed to the theory that the mass extinctions were due to gradual climate change. The dinosaurs vanished just as the shallow inland sea that covered much of the midwestern United States receded, probably due to a drop in the global sea level. Such massive changes in the distribution of water over the Earth must have affected the wind and ocean currents, and thereby the worldwide climate. Isotopic measurements indicate that the global temperature did drop significantly at this time. With so many complex changes occurring on the Earth, it looked almost hopeless to try to explain the extinctions from a single cause. Paleontologist Steven Stanley says in his text *Principles of*

Paleontology: "One conclusion that we will always be safe in drawing is that the causes of mass extinction are not simple."

Scientists are trained to be skeptical, to doubt, to test everything. But when people talk about the "scientific method," they never mention that too much skepticism can be just as bad as too little. When presented with a new, startling, and strange result, it is easy to find flaws and come up with reasons to dismiss the finding. Even if the skeptic can't find an outright mistake, he can say, "I'm not convinced." In fact, most scientists (myself included) have found that if you dismiss out of hand all claims of great new discoveries, you will be right 95% of the time. So skepticism is rewarded. But every once in a while, there will be that rare occasion when you are wrong, and you may have missed an important discovery because of it.

Likewise, you cannot afford to lose your skepticism or you will waste your time in hopelessly blind alleys. How do you develop the right sense of skepticism—when to dismiss and when to take seriously? Luie's sense of skepticism was clearly finely tuned. Those who dismissed his discoveries prematurely would be doomed to continue their current line of research. By the time they finally accepted Luie's work, all the important new discoveries would have been made. How do you argue with someone who has a different level of skepticism? How do you respond to the statement "I am not convinced"? The best way, the only possible way, is to go on with the work. Be grateful that the competition has not even entered the race, and has left all the fun to you.

Luie should be happy, I thought. He had solved one of the great mysteries of science. *I* knew he was right. No other theory adequately explained the iridium. And yet in the fifteen years that I had known him, I had never seen him so unhappy. He could simply wait for history to make the judgment and prove him right; yet he took a great deal of time and trouble to respond to all the papers criticizing his theory. He sometimes was so furious at some of the papers and letters attacking his model that his complexion noticeably reddened. Old theories never die, I thought, only old theorists. Eventually the present generation of paleontologists would go away and be replaced by a new group, young students not tied to the old ways. That was the history of most new theories. I advised Luie to be patient.

Then I realized that most of the paleontologists who disputed Luie's discovery would outlive him. Luie was over seventy years old. I could afford to be patient, but Luie couldn't, not if he was going to see his theory become part of the standard dogma. His father had lived well into his nineties, and I had always expected Luie to live forever. But perhaps his impatience was understandable.

Meanwhile, Walt kept collecting rocks, and Frank and Helen kept analyzing them. The group began to attract collaborators, geologists who were happy to send samples to the Lawrence Berkeley Laboratory for analysis. The impact theory had many implications to be tested. The iridium enhancement should be found worldwide. By 1986 it had been found in more than eighty locations around the world. The theory had predicted that the clay should have a similar composition everywhere. It did, to the extent that that can be tested after 65 million years. (Chemical processes under the ocean continue to alter the rock.) The chemical composition of the clay from Denmark and from the north central Pacific agreed as closely as two clays from sites 1 kilometer apart in Denmark, and were chemically dissimilar from the clay entrained in the limestone above and below the boundary. This was just what the impact theory had predicted, since the boundary clay layer worldwide came from the same impact crater, whereas the entrained clay in the limestone came from local rivers.

Walt suggested that the group look for another iridium level, at the Eocene-Oligocene (E-O) boundary. This boundary marked the time of a mass extinction that occurred on the continents between 35 and 39 million years ago, at the end of the Eocene. Since this boundary was not as well defined as the Cretaceous-Tertiary (it had no obvious clay layer), Walt suggested that they look near a region where microtektites had been found.

Microtektites were thought to be droplets of glass formed in the splash of a large impact. Their occurrence near the E-O boundary had been carefully mapped by the propitiously named scientist Billy Glass. Soon Frank and Helen found the iridium layer, the second ever discovered, exactly coincident with the microtektite layer. The iridium at the E-O boundary was discovered independently by R. Ganapathy, a scientist who had managed to continue his research on meteoritics while working for an industrial company.

"Why hadn't the lack of sunlight during the Cretaceous killed the plants?" asked some of the skeptics. It was a difficult point to answer, since photosynthesis would stop in the darkness. Leo Hickey, a paleo-botanist who had been a graduate student at Princeton with Walt, chided,

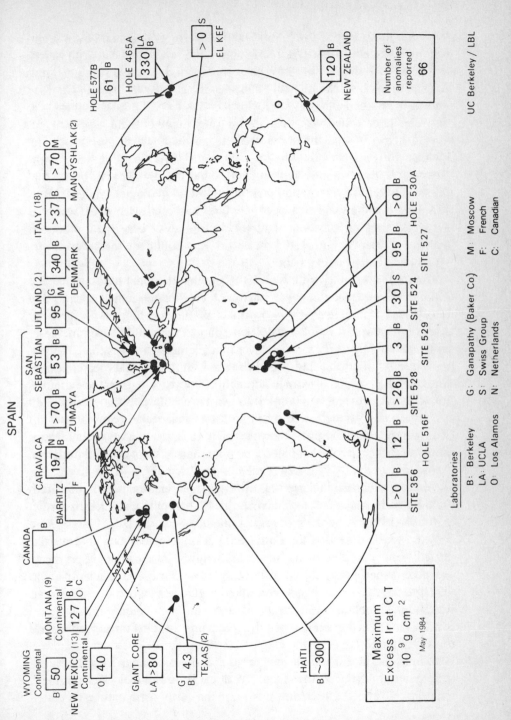

Locations around the world where the clay layer has been found, as of May 1984.

in several professional articles and talks, that the plants didn't even seem to *notice* the catastrophe. Of course, plant life is very robust, with much of it accustomed to living through severe periods such as droughts and freezes. In my own experience, I had seen the eucalyptus trees of California killed by a short frost, only to spring back from the surviving roots a few years later. Since plants cannot migrate, many of them have evolved ways to leave behind the seeds of their survival, which are able to lie dormant for long periods. They are, in many ways, ideal organisms for surviving a three-month dark period. After the light came back, the plants should have recovered quickly, at least by geologic standards. But shouldn't there be *some* evidence that plants had been affected? It was one of several nagging questions that wouldn't go away. It seemed to bother Walt more than it bothered Luie, who was confident that everything would eventually be understood.

The most annoying critic was Bill Clemens, the local dinosaur expert in the Berkeley Paleontology Department. At first he seemed open to the theory, and even collected rock samples at sites in Montana for neutron activation analysis. Frank and Helen found the iridium layer in these samples, but they were in a rock layer that Clemens said was three meters *above* the last dinosaur find. According to him, that would mean that the dinosaurs had become extinct about 30,000 years *before* the putative impact had occurred, a potentially devastating blow to the theory. His conclusion was strengthened when he found Paleocene mammals below (i.e., before) the iridium layer. He claimed that it was known from numerous other fossils that Paleocene mammals had not evolved until the dinosaurs had totally vanished. Luie felt that Clemens must be wrong. He looked at Clemens's evidence that the Paleocene mammals had postdated the dinosaurs and was "not convinced." The trouble with skepticism is that it works for either side of an argument.

Walt preached caution for a different reason. Geology is a notoriously difficult and complex business. Walt delights in showing others how complex. When I visited him in Italy, he showed me a dark layer of rock called the "Bonarelli." Then we walked a hundred yards, to a place that was about 20 feet higher in the stratigraphy, and, to my amazement, there it was again! Farther on up the hill, it appeared a third time. Were there three Bonarellis?

"This was once a submarine slope, and the sediments were likely to slide down it during earthquakes," Walt explained to me, "so right here the Bonarelli appears three times, twice in the correct orientation, repeated by submarine landsliding." One had to be very careful to be able to tell which layers had been laid down first.

Luie claimed publicly that the 30,000-year gap between the last articulated dinosaur bone (one that clearly hadn't moved since it was put down) and the iridium level was insignificant, particularly when compared to a 65-million-year time span. "That's less than a part in a thousand. Pretty good agreement," he said. He was sure that the gap would go away as more precise measurements were made. The paleontologists thought Luie's dismissal of their data somewhat cavalier. Some of them were bothered by the fact that he considered only the articulated bones as worthy of study. Luie argued that his choice was based on his vast experience with data analysis. "You have to learn to recognize bad data," he said. The paleontologist, given this choice, would have liked to dismiss the iridium data. That was in a different category, and it couldn't be dismissed, according to Luie. Luie's position seemed to me to be unfair. If you are allowed to throw away any data that you disagree with, then you can reach any conclusion you want. Luie listened to these complaints patiently, but insisted that he was right. To those who disagreed (and to some who agreed) it seemed as if Luie was arrogantly "pulling authority."

Not all the paleontologists rejected the impact theory. Dale Russell, an internationally known Canadian paleontologist, was thrilled by the work at Berkeley and decided to spend a year's sabbatical with the group. He had recently made some interesting discoveries concerning the dinosaurs' intelligence. He had found one dinosaur, the stenonychosaur, that appeared to have a higher brain-to-body weight than that of any other creature that had existed up to that time, mammals included. Evolution had just begun to discover the advantages of intelligence, and it was evolving rapidly for the dinosaurs. Had the catastrophe not taken place, the dinosaurs might have evolved to be now as intelligent as we are, or more intelligent. Russell helped produce a model that shows what the dinosaurs might have become had they not been destroyed. With large humanlike heads but strong reptilian features, the images are unforgettable, and they make the point more effectively than any technical explanation. Dale once told me that he felt the evolution of intelligence may have been set back millions of years when the dinosaurs were destroyed. Dinosaurs were certainly not an evolutionary dead end.

There is an analogy between the development of intelligence and the development of computers. Until intelligence developed, the programming of animals (i.e., their behavioral rules) was all "hard-wired," like a special- or single-purpose computer, which could do the task it was designed for and nothing else. If Nature wanted to try a new idea, it had to develop a new special-purpose computer. That was what Nature did throughout most of evolution. It was a slow process, particularly since the method was trial-and-error. With the advent of intelligence, Nature had a general-purpose computer. Changes could be put in the software (memory, language, culture), rather than in the hardware (instinct, the central nervous system). Previously, voice had been useful for sending signals, such as warnings, comparable in this analogy to computer interrupts. But with high intelligence, language was developed that was complex enough to transfer programs (instructions, advice, stories).

Russell understood Luie's impact-theory calculations and examined the consequences for fossil data. To the surprise of many of his colleagues, he concluded that the extinction of the dinosaurs may indeed have been abrupt. Dinosaur fossils are rare, and one cannot take the last appearance of a particular species as the time when it became extinct. Since the chance of finding any one skeleton of a rare species is so low, you shouldn't expect to find one right at the boundary. Russell showed that the average vertical spacing between dinosaur fossils in Montana was about 1 meter. From this one could use Poisson statistics to calculate the probability of getting a 3-meter gap, and the answer was $e^{-3} = 0.05 = 5\%$, or 1 out of 20. These weren't bad odds. Russell concluded that the examples of premature dinosaur extinctions could all be explained as statistical and systematic artifacts.

To get a better measure of the abruptness of the mass extinction, Luie said, we should use abundant fossils, such as the forams, which had an average vertical spacing of less than a millimeter. Their demise offered a far more accurate measure of when the catastrophe occurred, and they had disappeared precisely at the iridium layer. Luie said that this example verified the abruptness of the event.

But Russell's mathematics couldn't explain the Paleocene mammals found below the iridium layer. Luie's dismissal of this fact seemed arbitrary and self-serving. Then, to my surprise, the conflict was resolved. A geologist named Jan Smit went with Bill Clemens to study the rock layering, the stratigraphy, of Clemens's Montana location. Clemens had assumed a "layer cake" stratigraphy, with all the rock layers neatly preserving the order in which they had been laid down. He had taken the rock samples for iridium analysis from a different location from the one in which he found the Paleocene mammals, and assumed that the vertical structure of the rock layering was the same in both locations. The geology turned out to be more complex than that. Smit found that Clemens had taken some of his samples in an area where "channeling" had eroded away part of the rock layers, and washed down rock and fossils from more recent times to replace them. Thus the Paleocene mammals were below the iridium layer only because they had been washed there after the layer had formed. Upon close inspection, it turned out that the Paleocene mammals were later than the iridium layer. Luie's instinct had turned out to be correct.

However, even with his most dramatic counterevidence proved wrong, Clemens dug in his heels. He refused to accept the idea that an impact had killed the dinosaurs. Luie was convinced that Clemens was a reasonable man, who could eventually be brought around; all it would take would be time and patience. A weekly seminar was set up to discuss the physical and paleontological evidence. Every Tuesday morning for three months, Luie met with Clemens, Russell, Walt, Frank, and Helen, and anybody else who wanted to attend. I never did, but Luie kept telling me his latest ideas for convincing Clemens. They never worked, and Luie never really understood why. I suspected that Clemens may have felt a similar frustration with Luie. Clemens refused to accept Luie's contention that the extinctions were abrupt, and Luie refused to accept Clemens's evidence that they were not. In retrospect, I believe that they were both right. The resolution of this paradox was still several years off, awaiting the Nemesis theory.

A few other paleontologists besides Dale Russell supported the Alvarez mechanism. Hans Thierstein, a Swiss paleontologist working at the Scripps Institution of Oceanography, in San Diego, had studied the microplankton, the microscopic creatures that were at the very bottom of the food chain in the oceans. These were the animals that would be most directly affected by the absence of sunlight, since the ecology becomes more and more complex and intertwined the higher up the food chain one looks. Thierstein concluded from his detailed studies of the plankton extinctions at the end of the Cretaceous period that "darkness is a very good mechanism that could account for the fossil plankton we see."

Clemens found a new reason to dispute the impact hypothesis. The rock samples collected in Montana had their remnant magnetism measured by Robert Butler, a geophysicist at the University of Arizona, and most of them showed that the Earth's magnetic field had been reversed (compared with its direction now) when the layer was formed. But in one section the rock showed a normal field, not reversed. That proved that the clay layer could not have been formed at the same time at all locations. It didn't explain what had formed the layer, but it certainly ruled out the possibility that the layer was the result of a single large impact.

Walt visited the Hell Creek site in Montana with Clemens, Shoemaker, Hickey, and Butler. As they collected samples around the area, Walt and Gene became suspicious. It was impossible to measure the remnant magnetism in the field, but there were other indications that the clay layer had been laid down all at one time. Hickey found and identified key fossils that indicated synchronicity. As the group discussed the local geology together, Butler also began to have doubts about his former conclusion. That evening he carefully reviewed all his old notes.

The next morning Butler told Clemens that he would like to address the entire group. Everyone gathered around as Butler climbed onto the fender of the jeep to make an announcement. When he reviewed his notes, Butler said, he had found a mistake. The previous rock samples had all been collected by Clemens and J. David Archibald, a paleontologist at San Diego State. Butler had measured the paleomagnetism at his lab in Arizona. But there had been a transcription error. The magnetic data and the rock samples had been shifted with respect to each other when the data were transcribed. When the error was corrected, the data showed that all the samples had the same magnetic polarity. The published results were wrong. The clay layer had been laid down at the same time everywhere.

Luie says that a scientist's willingness honestly and openly to retract his mistakes is more important in establishing his scientific stature than a long list of discoveries. One doesn't usually insist on their making the retraction immediately, and in such a dramatic fashion, however. Walt was also impressed with Butler's candid announcement of error. "A lesser scientist," he said, "would have tried to cover up the mistake, or maybe argue that he had underestimated the error uncertainties previously, but Butler just came out and announced to everybody that he was wrong." It was a proud moment for science.

Another wrong piece had been removed from the jigsaw puzzle. It is rare for a theory to be so good that it can be used to predict that certain measurements must have been done incorrectly. The impact theory had this virtue, and it greatly strengthened the confidence of Luie's team. The flow all seemed to be in one direction, toward proving the impact theory right. Contrary evidence never seemed to last more than a few months, as evidence in favor of the impact steadily built.

The cores of sedimentary rock that I had seen in the basement at Lamont were analyzed, and the iridium-bearing layer found. As the number of locations around the world where the layer was found grew to more than eighty, nine different groups in seven countries became responsible for the measurements. The iridium had been laid down world-wide. The amounts at different locations did not always agree, but there was always a significant excess. Yet early hopes that the pattern could be used to help locate the impact site proved fruitless.

Other theories were published, which, their authors claimed, could account for the iridium. Iridium is present in the ocean at all times, one investigator pointed out. A shift in the chemistry of the oceans could cause the iridium to precipitate out of solution, forming salts that then settled to the ocean floor. Walt did not consider this theory credible, because it seemed unlikely that all the world's oceans could change at the same time. But it was one thing to convince yourself and another thing to convince a skeptic. Walt could not prove that such a precipitation had not taken place.

To refute the model that said the ocean was the source of the iridium, it would be sufficient to find the iridium in sediment laid down in the middle of a continent. Walt applied to the National Science Foundation (NSF) for funding to locate and take samples from a suitable lake site. The peer reviewers, anonymous scientists working in related fields, said the money would be wasted, since the source of the iridium was obviously from precipitation in the oceans. You cannot find iridium if it is not there. Walt's request was turned down. He and Luie were outraged. They could have appealed the NSF denial, but that was hardly worth doing considering the bad referee reports. Even if they were to win the appeal, it would take time, perhaps a year. The peer-review system is a very conservative system, and is effective at filtering out the "nut" proposals, even those few that turn out to be right.

One day Luie came into my office very excited by a preprint that he had just received from Carl Orth, Charles Pillmore, Robert Tschudy, and their group at the U.S. Geological Survey in Denver. "This paper is a classic!" he exclaimed. "This diagram will someday be reproduced in every text-book on geology or paleontology. Look at this!" The diagram plotted the density of iridium as a function of depth, with a sharp increase (by a factor of 300) right at the level identified as the Cretaceous boundary. And right next to it was a plot that showed the ratio of pollen to fern spores. This ratio dropped abruptly exactly at the iridium level. Luie said, "This proves the plants were affected at exactly the same time as the animals. But what's more, their samples were taken from a continental deposit!" That completely ruled out chemical precipitation in the oceans as a source of the iridium. Luie and Walt weren't surprised, since they had already concluded that the iridium must have been extraterrestrial, but they were pleased to see such a clean new result prove them right.

Some scientists had claimed that a volcanic eruption could be the source of the iridium. Lava from volcanoes comes from deep within the Earth, so couldn't it be enriched? The Alvarez team had considered this possibility and answered it to their own satisfaction. First, numerous measurements had shown that the lava from eruptions is not enriched in iridium. In fact, Frank could consistently identify volcanic layers by the absence of iridium. Since the layers had been set down rapidly, they did not contain much trapped iridium from micrometeorite dust. Second, the ratios of abundances of other elements, particularly gold and platinum, were considerably different from ratios in the Earth's crust and in lavas from the mantle. However, some critics did not accept these arguments. One volcanic eruption of Kilauea had shown an excess of iridium in the vapors it emitted (although not in its lava), but Walt pointed out that Kilauea was right on the boundary of a continental shelf, and undoubtedly had roots deep in the mantle. Such volcanoes have relatively gentle eruptions, more like leaks than explosions, and thus they could not have created the world-wide clay layer.

Meanwhile, evidence favoring the impact theory was accumulating. Miriam Kastner, a geochemist at the Scripps Institution of Oceanography, studied the clay layer in Denmark and in the north central Pacific, and showed that it was composed largely of smectite, an alteration product of glass that was much more likely to have come from an impact than from a volcano. Jan Smit discovered millimeter-size mineral spherules in the clay, which were shown by Alessandro Montanari, a student of Walt's, to be the fossilized remains of glassy microtektites formed in the splash of molten rock in the impact.

Karl Turekian was the most eminent of those skeptics who felt that volcanic eruption was a possible alternative to the impact theory. Turekian, a geochemist at Yale, took on the task of deciding between the two possibilities. He knew that extraterrestrial material and volcanic rock could be distinguished by other ways than iridium, and he decided that the most sensitive would be by using the ratios of isotopes of the element osmium. Turekian and his group made the measurements themselves and concluded that the material found in the layer had not come from the Earth's crust, but had a significant fraction with an extraterrestrial origin. Walt felt that Turekian's support represented a turning point in acceptance of the theory. Turekian was the first major geophysicist to change his mind publicly, and to say that the impact claim must be taken seriously.

A few weeks later, the Berkeley group was surprised to receive a preprint of a new paper from Bruce Bohor and his collaborators at the U.S. Geological Survey. They had analyzed minute crystals of quartz that had been found in the clay layer, and discovered that the quartz had a layered structure that indicated it had been subjected to a severe shock. Similar layering had been seen in quartz crystals adjacent to Meteor Crater and in craters formed by nuclear-bomb tests, sites of exceedingly high-pressure shock waves. Pressures able to cause such shocks could never build up in a volcanic eruption, because the Earth would yield and release it. The shocked quartz ruled out the volcanic theory completely.

But the volcanoes continued to get attention, largely because of a set of papers by Charles Officer and Charles Drake of Dartmouth College. In their first paper they outlined all the evidence in favor of volcanic origin. The Berkeley group was upset by this paper, not because of its arguments, but because they felt it ignored the possible counterarguments and gave insufficient attention to the quality of the data they cited. For example, Officer and Drake argued that the clay layer was not laid down simultaneously all around the world. In support of this, they cited the work that Butler had repudiated on his jeep. For this they could be forgiven; they undoubtedly had not heard about the error. But they also cited evidence from measurements made of seafloor cores. When Walt checked the references, he found that, in the original articles, the authors called their own data the "worst" they had seen and said that it was not to be "trusted." Officer and Drake gave the impression in their paper that iridium from volcanoes is common, when in fact it is virtually absent.

The Berkeley group wrote a detailed response to the Officer and Drake paper, rebutting it point by point. They also summarized the evidence that ruled out the volcanic origin, which now included the shocked quartz, the spherules, the osmium isotope ratios, the smectite, the lack of iridium in volcanic eruptions, and the observation that the ratios of the various siderophile elements found in the clay layer matched those in meteorites, rather than the ratios found in crystal rocks.

Luie expected Officer and Drake to concede, so he was surprised when *Science* published a new paper, which referred to their rebuttal but didn't answer any of its objections in any detail. Instead, Officer and Drake simply stated, "Despite the various criticisms, Officer and Drake stand by their original arguments." Other rebuttals to Officer and Drake, pointing out misinterpretations and errors, were published by Smit, geophysicist Frank Kyte, and Bevan French, a geologist at the Goddard Space Flight Center.

The editors of the *New York Times* uncritically accepted the conclusions of the Officer and Drake paper and, in an unusual action, stated on their editorial page that the impact theory was dead. This pronouncement caused some immediate concern among the funding agencies in Washington, and letters had to be sent to them immediately saying the case was not as closed as the editorial indicated. I was amazed that the *New York Times* would publish an editorial without checking sources for accuracy, particularly since the science writers for the newspaper were generally good. I called one of them to find out how they could have let such a thing be printed and learned that the science department had had no advance warning. They were just as surprised as we were. An editor of the paper conceded to me that they don't check their facts, since editorials are opinion, not fact.

In December 1985, Officer presented a paper at the annual meeting of the American Geophysical Union in San Francisco. As I listened in the audience, I heard him repeat many of the claims that the Alvarez group had already rebutted in print, but he had at least one point that had not yet been publicly answered. He reiterated a claim that had been made in his second paper with Drake. He said that the "spherules," the so-called fossilized microtektites that the Alvarez group had used as evidence of impact, were not unique to the Cretaceous boundary. He showed several slides of little balls found by an Italian group, which he said were identical to the Alvarez spherules. These balls had been found in all the rock above and below the clay boundary. He concluded that it was wrong to interpret the spherules as evidence for an impact. They probably came from volcanic eruptions. It must have appeared to the audience that the Alvarez group had in fact been very sloppy in their collection of the original samples.

In the question period that followed Officer's talk, Walt was the first to be recognized by the moderator. Walt had decided not to water down his response by replying to all the old points, the ones his group had answered in print, but, instead, to spend his allotted two minutes addressing the new claim about the spherules being everywhere.

He explained that at first he had been very concerned about the published discovery of spherulelike "little balls" in rock above and below the Cretaceous boundary, since this find was obviously contrary to his own work. Walt's student Alessandro Montanari had therefore looked for them at their site in Italy. Sandro had no difficulty in finding the little balls in virtually all the rock above and below the boundary clay, but these little balls didn't look to him like the true spherules they had found earlier. That's why they had been ignored by the Alvarez team. Looking at them under a microscope, Sandro had discovered they were hollow, unlike the spherules, but just like the little balls found by Officer and Drake. What were they? He tested them by applying hydrofluoric acid. True spherules are made of sanidine and would dissolve, but the little balls didn't. He poked them with a needle; they flexed. He guessed that they were made of organic, not mineral, material. Sandro then heated a few of them, and they burned! Finally, Walt explained at the conference, Sandro took them to a biologist, who readily identified the little balls as modern insect eggs.

At this point in Walt's narrative, the audience of over a thousand scientists roared with laughter. Walt continued: Sandro had found that these insect eggs abound on all surface rock, above, below, and at the Cretaceous-Tertiary boundary. They were what Officer and Drake were looking at. But, *if you are careful* and dig into the rock to avoid surface contamination, they are missing. Deep in the rock you don't find any insect eggs, only true spherules, and these you find only in the boundary clay.

Part 2

NEMESIS

9. Blind Alleys

I HAD watched the revolution from the sidelines, with a suppressed sense of jealousy. Luie and Walt had opened a new field of science, and I had missed out, just as Luie had missed out on the discoveries of fission and the chain reaction. I couldn't join their team unless I had something to offer, and I had nothing to offer. With Luie spending full time thinking and analyzing, it was hard enough just to keep up.

Then, in late 1983, Luie received the "crazy paper" from paleontologists Dave Raup and Jack Sepkoski in which they claimed that the Cretaceous catastrophe and other mass extinctions took place on a regular schedule, every 26 million years. Although Luie often embraced crazy ideas, he dismissed this one. It was a slip by the master of balanced skepticism. Perhaps he had become too caught up in his own theory. Perhaps the fame of his recent discoveries had brought him so many nut letters that he didn't have the time to look at them carefully. He asked me, as I said earlier, to play the role of devil's advocate, to take the side of Raup and Sepkoski, and I did. In the process I suggested that their data could be understood if there was a star orbiting the sun. Luie's dismissal of the new paper had suddenly thrust me into a key position, and before I knew it I was in the midst of the controversy.

I originally suggested the companion-star idea as a debating point, to show Luie that his logic was flawed when he dismissed some of the Raup and Sepkoski data. I didn't consider it a serious scientific contribution, and so I was bemused and aloof when he immediately said that he wanted to call Raup at the University of Chicago and tell him that I had found an explanation for their data. Raup was away, but someone took a message. Alvarez said he would let me know as soon as Raup got back to him, so we could tell Raup about the idea together. "Don't bother," I diffidently replied. "I'm sure you can explain it to him without my help."

A short while later, Luie came to my office. "Raup just called," he said. "I told him that my young colleague Rich Muller had invented a model that could explain his periodicity. I told him about your idea. Raup said that someone else had invented the solar-companion-star idea last week. Raup told Shoemaker about it, and Gene said the orbit of the star would be unstable. Your idea doesn't work."

Luie believed that Gene Shoemaker knew more about solar-system dynamics than anybody else in the world, and I had to take his criticism seriously. What did it mean for an orbit to be unstable? I guessed that he was referring to the gravitational effects of nearby stars, which could gradually distort the orbit of the companion star until it became free, left the sun, and sailed off into space.

A short while later, fellow physics professor Frank Crawford came by my office, and I told him about the interesting events of the previous hour. He loved the thought that the sun could be a binary star. If Shoemaker was right, he wouldn't have the pleasure of thinking about the consequences, so he assumed that Shoemaker had made a mistake and started playing with the idea himself.

"Say, how do we know that Alpha Centauri doesn't orbit the sun?" Frank suddenly asked. Alpha Centauri is the closest known star system to the sun. The name comes from the fact that it is the brightest star (hence alpha, the first letter in the Greek alphabet) in the constellation Centaurus, the "centaur."

"Alpha Centauri is too far away," I immediately replied. "The companion has to be less than three light-years away, and Alpha Centauri is over four."

Frank asked to see my calculation of the size of the companion's orbit. He immediately spotted an error. "You neglected the mass of the companion star," Frank said. "If the star is massive, it could be farther away. What is the mass of Alpha Centauri?"

After a few minutes' search through my cluttered office, I found my copy of C. W. Allen's *Astrophysical Quantities*, a thin green book full of astronomical statistics that every astronomer tries to keep nearby. The mass of Alpha Centauri was listed as 1.1 solar masses. That wasn't big enough. For something as far away as Alpha Centauri to orbit the sun with a 26-million-year period it would have to weigh 2 solar masses. Frank's Alpha Centauri idea didn't work.

A few minutes later, Frank came back into my office and asked, "How do you suppose they know the mass of Alpha Centauri? You can measure the mass of stars accurately only if they are part of multiple-star systems. Alpha Centauri must have a companion." It was not the mass of Alpha Centauri that was important; it was the mass of the Alpha Centauri *system*.

"Of course! You're right!" I said. "Proxima Centauri orbits Alpha Centauri." It was a fact I had learned as a teenager, when I built a 6-inch telescope to try to see the stars through the haze of the Bronx sky. Proxima, meaning "near," was named that because it is actually a little closer to us than Alpha is. We went back to Allen's book, which showed that there were three stars in the Alpha Centauri star system. "Not including the sun," Frank joked. The masses added up to almost exactly 2 solar masses. That was just what was needed to give the system the required 26-million-year orbit around the sun. We couldn't believe our eyes. We have either made the greatest discovery of the century, I thought, or we are doing something incredibly stupid. Since Alpha Centauri is one of the brightest stars in the sky and is famous for its nearness, how could astronomers not have realized that it was orbiting the sun?

It turned out that we *were* doing something stupid. A little while later I noticed the listing for "proper motion" in the table of nearby stars, with a value of 3.68 seconds of arc per year for Alpha Centauri. This meant that it was moving past us at a high velocity, more than 20 kilometers per second, faster than the escape velocity for that distance. It was not orbiting the sun, but was just passing by. The fact that the mass had turned out to be 2 solar masses had been just a coincidence. Well, you have to have lots of ideas that don't work in order to earn the one that does. I had enjoyed that half hour of thinking we had made a great discovery.

The next day I amused Luie by telling him about our Alpha Centauri fiasco. He asked me whether I had figured out yet why the orbit was unstable. I told him I hadn't. Actually, I hadn't even thought about it. Obviously, Crawford wasn't the only one who thought I shouldn't simply accept Shoemaker's expertise. Even though Luie was so skeptical of Raup and Sepkoski's work, he knew I wasn't, and he expected me to follow up on my idea of a solar companion. He knew how easy it was to be lazy and not follow up on a new idea. He wanted to make certain that I didn't make that mistake.

After this prodding by Luie, I went down to the Physics Library and took out several books on orbital dynamics. They were full of calculations of perturbations of orbits. I found the jargon almost impenetrable. It seemed as if the people who wrote these books were addressing only other experts. I guessed that almost nobody but orbital dynamicists ever picked up books on orbital dynamics. It looked hopeless unless I took a few months to dive into the dense material. I really didn't want to do that.

It was hard for me to take my own idea too seriously in part because its consequences would be too fantastic. If there *was* a star orbiting the sun, we would have to change our entire theory of the origin of the solar system. The star would have been a decisive factor in the evolution of life, due to its periodic shotgunning of asteroids at the Earth. The asteroid that killed the dinosaurs had been the bullet, but the solar companion star was the murderer. The very revolutionary nature of the idea made me think that it was unlikely to be true.

I took the books on orbit calculations home one evening, to try once again to make some sense out of them. Maybe, somewhere, buried deep in one of them, the author would say something like: "Large eccentric orbits are unstable due to the effects of passing stars, as can easily be understood from the following simple arguments. . . ." But, alas, I could find no such passage. I went to bed, but couldn't stop thinking about the problem. I visualized a long eccentric orbit, with one end going extremely close to the sun. The gravity from a nearby star was pulling on the Earth-companion system. I could imagine it near the companion, pulling slightly harder on it than on the sun, twisting the system. Twisting it! That was it, I suddenly realized. The twist would impart angular momentum to the system, and the distance of closest approach is proportional to the square of the angular momentum. It could go close to the sun one time, at most, before it picked up too much angular momentum. On its second orbit it would miss the sun by a huge distance, too far to cause a second extinction. The details could wait until morning. But I couldn't sleep. My mind kept racing. I got up and made a quick calculation with the help of pencil and paper. On the second orbit the companion star would miss the sun by tens of billions of miles, over a hundred astronomical units. Shoemaker was right.

The next day Luie once again asked me whether I understood the instability. This time I had been waiting for the question. "Yes!" I said, and gave him the simple argument from the previous night. He seemed impressed. The reasoning was so simple, so clear, that it was obviously right. The orbit was unstable, and my companion-star "theory" didn't work.

Although he still wasn't convinced that the Raup and Sepkoski data were correct, he knew that I took periodicity seriously, and he expected that I would keep trying to invent new models until I found one that worked. That's what he would have done. So every day he asked me if I had any new model to explain the periodicity.

To make the orbit stable I could not allow it to come close to the sun. Was there some way that it could miss by a large distance and yet make an asteroid hit the Earth? Most of the asteroids that orbit the sun are in a "belt" between the orbits of Mars and Jupiter. Maybe there was some way that a relatively distant passage of a companion star could perturb the belt, forcing some fraction of the asteroids into new orbits, orbits that would bring them into the path of the Earth. Once perturbed into such orbits, they might stay there for a long time. The Earth, for a few million years, would be in the midst of an asteroid storm. Whenever the Earth passed through the orbit of an asteroid there would be a slight chance of being hit. The probability was roughly the ratio of the diameter of the Earth (8,000 miles) to the length of the Earth's orbit (600 million miles), or 1 in 75,000. That meant that a collision would be very likely by the time 75,000 years had passed for each asteroid whose orbit intersected that of the Earth. That is just an instant in geological time. To complete the model, I just had to find a way to get a distant pass of a companion star to perturb the asteroids. It was a good start to a new theory.

Unfortunately, it seemed impossible to solve the second half of this problem. The asteroids made many trips around the sun while the companion star passed by, and the effect on their orbits seemed to be the opposite of what I wanted. The asteroid orbits tended to become circular, and to stay away from the Earth. The asteroids could be thrown into the Earth's path only if the companion star came very close to the sun, but then the companion orbit would be unstable.

I decided that I wasn't making much progress, and I should talk to someone who really knew something about astronomy. I had never even taken an astronomy course (or a geology course, for that matter), and there was a lot I didn't know. I immediately thought of Marc Davis. Marc had received his Ph.D. at Princeton, where he worked with cosmologist P. J. E. Peebles; he had a very strong background in theoretical astrophysics. In fact, I had difficulty in understanding many of his papers, because they were at a higher level of mathematical sophistication than I was accustomed to. Davis had spent a few years at Harvard, and I had first met him there. At Harvard he had worked on a project to measure the motions of thousands of galaxies in an attempt to understand the very large-scale structure of the Universe. It had been an ambitious project and it had paid off when the group found that the random motions of galaxies were much larger than had been previously supposed. I had been one of the first people to suggest to Marc that he come to Berkeley, and I had been delighted when he accepted a faculty position at U.C. I felt that he had a lot to teach me. Now I had finally found an excuse to try to strike up a collaboration.

The next day I went down from the hill on which the Lawrence Berkeley Laboratory was perched, to Marc's office on campus. I showed him the Raup and Sepkoski data, with the 26-million-year extinction period, and reviewed some of the recent findings of Luie and his group on asteroid impacts. Then I told him of my failed attempts to find an explanation. I carefully did not tell him about the mistake Frank and I had made with Alpha Centauri, an incident I had described without qualms to Luie. I didn't know Marc well enough yet to be candid about how stupid we had been.

After I finished my story, Marc sat back in his chair and thought out loud. He didn't question the Raup and Sepkoski extinction periodicity; he was willing to accept my word on that. He launched right into the astronomy questions. "Well, let's see. We need a 26-million-year period. In astronomy we have so many phenomena that we can supply a period for any need. I think the thing that comes closest to having a 26-million-year period is the oscillation of the sun up and down in the Milky Way."

I knew I had come to the right person. It was an approach totally different from mine, and one that had led Marc in a direction that I hadn't considered. The Milky Way, consisting of all the stars we see with the unaided eye at night, is shaped roughly like a flat dish, with the sun about two thirds of the way out to the edge. From our viewpoint within the plate, the bright strip in the sky we call the "Milky Way" is what we see as we look through the plane of the dish, where most of the stars are, and the dark region of the night sky is what we see when we look out perpendicular to the plate.

The sun is not at rest in this system, however. It moves in a rough circle, orbiting the center of the dish. While it is doing so, it also bobs up and down slightly, like a horse on a merry-go-round. Every time its velocity carries it above or below the galactic plane, the gravitational attraction of the stars left behind in the plane pulls it back. It was this bobbing that Marc thought was relevant. He pulled out a book on galactic dynamics that he kept near his desk, and looked up the number. He found that the sun passes through the thickest part of the dish every 33 million years.

Thirty-three million years is not exactly the same as 26 million years, but it was close enough, we hoped. Besides, maybe the astronomers were wrong about the 33. It differed from the desired answer by only 7 million years, 20%, and errors this large are common in astronomy (and many other sciences, too). And if the astronomers weren't wrong, maybe the paleontologists were. Raup and Sepkoski had found a second significant periodicity in their data at about 30 million years. We decided not to worry about the difference between 33 and 26, not just yet anyway.

Marc had suggested a basic phenomenon, the bobbing up and down in the galaxy, and now we had to figure out a way that the bobbing would cause an impact on the Earth. We knew that there is a lot of matter in the galaxy that is dark, unseen. The dark matter had been discovered indirectly by its gravitational effects on stars. But astronomers knew very little about this "missing matter." So Marc and I felt free to postulate that it was this unseen matter that we were running into every 26 (or 33?) million years. We agreed that the idea was worth pursuing, and that we would both keep thinking.

At home I made some simple calculations about the motion of the sun in the galaxy. The farther above the galactic plane the sun moved, the stronger would be the force pulling it back. Its motion would be similar to that of a pendulum of a giant clock, ticking back and forth every 26 million years. I knew about how much mass there must be in the dark matter of the galaxy, about 50% as much as there was in visible stars. Suppose all of the mass was in the form of small asteroids, 5 miles in diameter, about the size of the one that killed the dinosaurs. What was the probability that the Earth would run into one or more on each passage through the galactic plane? This was something I could calculate. Once again I pulled out Allen's *Astrophysical Quantities*. The average spacing between stars near the sun is about 6.5 light-years. The galactic plane has about 0.02 grams of matter for every square centimeter of area, when looking down on it from above. A 5-mile asteroid would weigh 5×10^{17}

grams (i.e., 5 with 17 zeros after it), or about 50 billion tons. To get the right number of grams per square centimeter would require a spacing between asteroids of about 5 billion centimeters, or about 30,000 miles. The Earth is 8,000 miles in diameter. So the chance of hitting one would be 8,000/30,000 = 0.27, about 1 chance in 4. That wasn't quite high enough, but it was close.

How could I make the probability higher? Suppose most of these "galactic" asteroids were smaller. We can hypothesize anything we want about their size, because they have never been observed. Such small objects wouldn't reflect enough light to be seen by even the biggest telescopes. If the asteroids averaged 1 mile in diameter, one fifth the size I had previously guessed, then there would have to be $5^3 = 125$ times as many of them to make up the missing mass. The asteroid from the Cretaceous extinctions was known to be bigger than that, but that could be an exception. After all, it is reasonable to assume that we would find the biggest one first. The average spacing between the asteroids (assuming they were all in the same plane) then would be the square root of 125 times smaller than the value of the 30,000 miles I had previously calculated, or about 3,000 miles. The Earth would be sure to run into at least one of these every time it passed through the galactic plane, since the spacing was smaller than the size of the Earth. The theory looked good.

But I began to worry about the stability of the layer of asteroids. As the sun passed through this plane, its gravity would give a kick to even distant asteroids, and they would begin to scatter. Other stars would do the same thing. Pretty soon the asteroids would be spread out as much as the stars. There would still be plenty of collisions, but they would occur at random times, not just at 26-million-year intervals. I called Marc, and before I could explain my concern he told me that he was worried about the stability of the asteroid layer. I told him that I had the same worry.

The next day I was browsing in Allen's book when I found an entry called "Sun's distance from the galactic plane." The distance listed was 8 parsecs, about 26 light-years. That is very, very small on the scale of galactic sizes. The sun is virtually *in* the galactic plane! If our theory was right, we should be running into an asteroid right about now. But there hadn't been an impact for more than 10 million years, according to Raup and Sepkoski. Something was wrong. Maybe the astronomers had made a mistake in estimating our position. They must have, if our new theory was to be right. I would have to read the basic papers and find their mistake.

That weekend I came across a story in the *New York Times* describing

Raup and Sepkoski's work and their claim for the 26-million-year periodicity in the mass extinctions. The next week the story was syndicated,
and I saw it in the *San Francisco Chronicle*. My advantage over the other
scientists in the world had suddenly vanished. Instead of being one of a
small group of people who knew about their work, I was now only one of
a huge number of scientists who knew. The newspaper article didn't show
their data, so there was no reason for the other scientists to be skeptical
about the periodicity. Thousands of physicists would now be working on
the problem. I had blown my lead. And, in fact, in Louisiana that very
morning two astronomers, Daniel Whitmire and Albert Jackson, had
indeed read the story, as I was to find out several months later.

I visited Marc in his office and told him that I had found another
difficulty with our already problematical theory: the fact that we were in
the plane of the Milky Way right now. I had tentatively concluded that the
astronomers must have mismeasured the position of the sun.

Marc immediately disagreed. He had known that the sun is roughly in
the galactic plane, but he hadn't realized that this position was incompat-
ible with the extinction data. Marc said that the astronomers could not
have made a mistake. He patiently explained to me the several different
methods that had been used to determine the sun's position. It had been
measured by counting how many stars are above and below us in the
galactic plane. The numbers turned out to be about equal, implying that
the sun was right in the middle. This had been done independently for
many different star types, and all the answers agreed. There was no room
for error. It was our theory, not the measurement of the sun's position in
the galaxy, that was wrong.

We had come close, but without success. How could we salvage the
theory? If the layer of asteroids was not *in* the plane, but up out of it, then
we would hit the layer at the right time. But what could hold the layer
there? Nothing could. Then I realized that if the asteroids were actually
nearby, moving up and down *with* us, we would cross them just as they
turned around in their motion. There would still be stability problems.
But then I stopped myself. This theory was simply too fantastic even for
me. I would have to hypothesize a mass of asteroids in an orbit through
the galaxy that almost matched, and was just ahead of, the orbit of the
sun. The theory had become as absurd as that of the cosmic terrorist who
carefully aimed a single asteroid at the Earth every 26 million years. It
was time to look in a new direction.

Something had happened to me without my realizing it. I had begun to make the problem of the periodic extinctions my own problem. It was no longer idle curiosity for me, but the center of my attention. A few months earlier I had assumed that the giant puzzle of the dinosaur extinctions had been solved completely by Luie and his team, with no role for me to play. Now, in effect, I had gradually come to believe that one edge of the jigsaw puzzle wasn't really an edge, but had a few pieces sticking out. There was more puzzle to be solved. It was a larger puzzle than anybody had realized. How much larger? It was impossible to tell, but it was larger.

So far I had made no obvious progress, but I wasn't too discouraged. Someone had once said, "Research is the process of going up alleys to see if they are blind." Each blind alley I had found so far had taught me something. I had learned new physics and astronomy, about orbital stability and galactic dynamics. It was frustrating, but fun.

I watched my two-year-old daughter, Melinda, learning how to walk up and down stairs. Accomplishing a new task, previously impossible for her, gave her so much joy that she would start giggling with excitement. Humans have an instinct that makes them enjoy learning. What a pity that so many of us slow down our rate of learning as we reach adulthood. I had learned more in the previous month, motivated by the periodicity problem, than I had learned in the previous year. It was exhilarating.

As I lay in bed one evening, trying to fall asleep, I kept imagining the sun bobbing up and down in the galactic plane. Once again my thoughts turned to the companion-star idea. What would be the effect of this bobbing on a companion star? The galactic tides could gradually remove angular momentum from the orbit. And they would also cause the orbit to precess, that is, change the direction of its axis. I visualized an elliptic orbit, gradually growing narrower, with the companion coming closer and closer to the sun on each orbit. Finally the companion would get so close that it would come right into the inner solar system, so close that it might be capable of triggering a disaster. *Disaster*. That word seems right. The word comes from Latin roots meaning something like "bad star." But then the star would move away. On its next orbit it wouldn't come close, since the narrow orbit is "unstable." The orbit would get fatter and fatter. The star would stay farther and farther away. And the axis of its orbit would move, swing, right past the galactic pole.

Suddenly I realized that if the axis of the orbit swung far enough, the force of the galactic tides would reverse. The orbit would begin to narrow again. The star would come back! Even though any particular orbit would be unstable, the whole motion over many orbits would be cyclic. Perhaps

with a 26-million-year period? I could use a relatively small (and therefore *stable* orbit) with a short period, maybe a million years. But the star
would come close to the sun only when the orbit narrowed sufficiently,
maybe every 26 orbits. That could give the 26-million-year disaster
period. It might actually work. I got out of bed, partly to write the idea
down (lest I forget it in the morning) and partly to see if I could work out
any of the equations. I couldn't. The only formulas I could find for
precession of orbits had to do with nearly circular orbits, not the eccentric
ones I was imagining. I went back to bed and tried to fall asleep again. No
luck. I couldn't get the wobbling orbit out of my mind.

The next day, after more frustrations with my calculations, I told two of
my graduate students, Jordin Kare and Saul Perlmutter, about the new
orbit I had imagined. They both offered to try to simulate the problem on
the computer. Meanwhile, I kept on trying to calculate. I looked at more
books, but as before they were of little help. It seemed that every problem
conceivable had already been worked out in these books except mine.
Several days passed. Finally I figured out a way to make some gross
approximations. The precession rate must be proportional to the galactic
gradient, as is the rate of change of angular momentum, the torque. I was
able to derive a rough formula, but it had one constant I couldn't calculate. I knew I would soon be seeing Freeman Dyson, a renowned theoretical physicist, at a meeting in Washington, D.C. I felt sure that he could
work out the answer in five minutes, on the back of an envelope. No, he
could probably work it out in his head.

Jordin was the first to get a computer simulation running, and he
plotted the orbit on the monitor. It was beautiful! The most amazing thing
was that it had worked just as I had predicted it would. I was surprised that
I was becoming such a good theorist. The oval orbit, oscillating back and
forth, first narrowing and then widening, only to narrow again, traced out
a path that looked like a tulip. I christened it the "tulip orbit."

I proudly showed the tulip picture to everyone who would look.
"What's that?" they asked, and my answer was always, cryptically,
"That's what killed the dinosaurs, maybe." Dyson's reaction when I
showed him the picture was, "That's very pretty." I asked him if he could
calculate the orbit for me, and he said he would think about it. It was not a
trivial problem, he said. I was disappointed. I always liked to think that
my theorist friends could calculate *anything*. Of course they always
seemed to assume that *I* knew everything there was to know about
experimental physics, so it was inevitable that we disappointed each

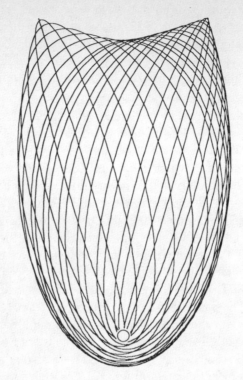

The tulip orbit.

other. Three days later, Dyson still had not solved the problem for me. He can be absentminded, and I suspected he had forgotten about it. Of course, it was also possible that even a great theoretical physicist couldn't solve every problem in three days.

Once again, I had trouble sleeping. I kept getting up to fiddle with pencil and paper. I'm not particularly good with elliptic integrals, but I was highly motivated. How could I get that constant? Suddenly I realized I could get it from Jordin's computer simulation. He couldn't get a formula, and I couldn't get the constant, but together we would have both. I had his plots at home with me, but I didn't know what scale he had used, so I had to wait until morning.

The next day I got to the laboratory early and waited anxiously for Jordin. When he finally arrived, I told him what I needed. It turned out that he had picked very simple constants for all the scales, to make the programming easy. I put the numbers into my formula and calculated the needed value; my constant turned out to be very close to the number 1. If I were really smart, I should have been able to derive that, I thought; in fact, a few years later I did. Now all I had to do was find an orbital period, eccentricity, and orientation that would give the required 26-million-year periodicity. With three values to pick, this shouldn't be too hard. Then I would have a theory!

The formula now showed that the catastrophe period was equal to the period of bobbing in the galactic plane, multiplied by itself, and divided by the orbital period of the companion star. The bobbing period was 33 million years. The orbit period had to be much less than that for the model to work, since the star had to make many orbits during a single bob. The implications of my algebra slowly became apparent. If I took 33 × 33 = 1,089 and divided it by a number much *smaller* than 33, the answer would always be much larger than 33. It would be *impossible* to find an orbit with a return period of 26 million years.

I reviewed my algebra again. I checked the constant I had from Jordin's program. The theory was beautiful. It was elegant. Tulip orbits were real, possible orbits, just as I had predicted. The orbits themselves were pretty, just as Dyson had said. The constant was 1. But the numbers didn't work out. That meant the theory was wrong. Dead. Tulip orbits might be interesting for some other problem, such as understanding the orbits of comets, but they couldn't give a 26-million-year-period mass-extinction cycle.

What really hurt was that the theory had been so beautiful. I had never thought that I had the ability to invent such a beautiful theory. My mathematics was too weak, I had always assumed. I was an experimentalist, not a theorist. But I had done it, created an elegant, beautiful theory—only to find it was irrelevant. A century earlier, Tolstoy wrote, "It is amazing how complete is the delusion that beauty is goodness."

At that point I really did stop thinking about the periodic extinctions. I had spent just over a month working hard on this problem, and had discovered that my best effort wasn't good enough. I wasn't sufficiently bright. My extraordinary effort had fallen short, so why bother trying any more? There was no way I could repeat, let alone surpass, the effort that led me to invent the tulip orbit. I would be wasting my time. Better to go back to work on my other projects. Maybe I could help on the supernova search. Christmas was approaching. It was time to get into the spirit. Rosemary and Betsy had noticed that I had become even more absent-minded than usual lately. I hadn't been indulging myself enough with the instinctive joys of playing with my children. I should spend more time dancing with Melinda to classical music. Try to teach her the concept of *grace*, just as I had taught Betsy several years earlier. I could now relax and enjoy life. Rosemary's father was visiting, and telling us the latest adventures he had had in the Canadian bush. Time to return to the real world. I took the family to see the *Nutcracker* ballet.

Marc didn't know I had given up, and on December 21, 1983, he phoned with some news. "Piet Hut is visiting," he said. "Piet is a *real* expert on orbital dynamics. We should talk to him about the extinction problem. He could be a big help." I momentarily considered telling Marc not to bother. And then part of me, the little smart part of me that sneaks out every now and then to surprise me with its excellent judgment, took control and said, "Sure." We decided to meet the next morning in Marc's office.

10. Eureka!

IT WAS the winter vacation, and the campus was almost empty the next morning as I drove my motorcycle to Campbell Hall, where Marc's office is located. Winter in Berkeley is often quite warm ("perpetual spring," I tell visitors), but this day was chilly, and the heating in the building was turned off.

In his office, Marc introduced me to Piet. I began by showing him the periodic-extinction data and describing all the failed theories we had made up in the last two months. He seemed surprisingly interested. I remembered my observation that to be a really good theorist you must find even *failed* theories interesting, for someday you may have to apply them in a new situation, where they might prove useful.

Piet was an excellent listener, and it didn't take long to go through all our work. I started with the original theory: The sun has a companion star with a very eccentric orbit, and the star comes close every 26 million years. I showed him my simple method of proving the orbit unstable. I surprised myself by telling him about the goof Crawford and I had made in thinking the companion might be the Alpha Centauri system. I told him about the two-step mechanism: The companion star in moderately eccentric orbit perturbs the asteroid belt. But the perturbation wasn't strong enough. I described the theory of galactic oscillations, and how we concluded that the sun is in the wrong place. And I described the tulip orbit. That was my grand finale. That was what Piet would like the most, I was sure.

He found the tulip orbit interesting, but the theory that he really liked was the idea of a companion star to the sun that went close enough to perturb only the asteroid belt. He thought that it was clever because it involved two disjointed ideas, two pieces of a puzzle that he might not have thought to fit together himself. First, the star comes close and does nothing drastic. It just messes up the asteroid orbits a little bit. But some of these messed-up orbits cross the Earth, and some of the asteroids hit. It lasts for only a few million years, at the end of which the perturbed asteroids are all destroyed by collisions or have inched back to more normal circular orbits. A very nice, two-step idea. Of course, it doesn't work, because the companion-star orbit is either too eccentric to be stable or too broad to affect the asteroids. But good theorists are just as interested in ideas that *don't* work, if they involve something new, something clever.

"And, of course, the star would have the same effect on the comet cloud," Piet added, as if it were the next thought that would occur to any half-wit. It had never occurred to me, or to Marc. It was a small change in the theory, but it was something we had missed. I started to get excited; so did Marc. Piet continued, "The star would scatter some of the comets toward the Earth, just as it scattered the asteroids. Could the impact that killed the dinosaurs have been a comet instead of an asteroid?" he asked me.

"Yes, of course!" I answered immediately. The possibility of a comet, instead of an asteroid, impact was something I had discussed with Luie several times. In fact, we knew more about the makeup of comets than of asteroids, since several meteor showers were known to have come from the breakup of comets, and we had pieces of the meteorites from them. The fact that extraterrestrial material was enriched in iridium had come from studying these fragments of comets. Which had it been, a comet or an asteroid? Luie felt the question was not important. But which was it? If he had to choose, he would refer to the experts. Shoemaker, the expert, said that more asteroids crossed the path of the Earth than did comets, so it was probably an asteroid. In all of the papers that Luie had helped write, it was called an asteroid. Comets were hardly mentioned, but we knew all along that it could have been a comet.

"Eureka!" I said it quietly to myself. It was the word Archimedes had shouted when, as he lowered himself into his bathtub, he saw that the level of the water was rising. He had suddenly found an explanation for why objects float: because they push water up. Legend says that he immediately jumped out of the bath and ran naked through the streets of ancient Syracuse in excitement at his sudden understanding of an old mystery. I wasn't sure enough of myself to run through Berkeley shouting aloud (let alone naked), but I sensed that Piet's modification eliminated all the obvious objections to the companion-star theory. The orbit for the companion star could be quite fat, only mildly eccentric, and it would still pass through the large comet region. That meant that the orbit would be stable, unlike the very narrow orbit that had been needed to perturb the asteroids. My days spent trying to understand orbital stability were finally paying off, as I found I could think about these new ideas quickly. Would the numbers work out? Were there enough comets out there?

I had never before regarded comets as particularly interesting. I knew that they were supposed to come from a distant region of the solar system, a region referred to as the "comet cloud." They are small objects, only a few miles across, but unlike asteroids they contain a lot of frozen water, ammonia, and methane. Dyson had once suggested that we might want to use them as a supply of water if we ever travel to the outer parts of the solar system. Some people speculate that comets hitting the Earth might have been the original source of water. When comets get close enough to the sun, the water and methane heat up and the expanding cloud of gases grows to an enormous ball, up to a million miles in diameter, called the "head" of the comet. Some of this head of gas is blown back away from the direction of the sun by the solar wind, and it creates a long tail tens of millions of miles long. The spectacular head and the even more spectacular tail create the beautiful sweeping image of the comet in the sky, and give the comet (which means "tail") its name. Most of what we see is thin gas. The tiny solid part of the comet, the core, is so small that one has not yet been seen. We know its size only from calculations, because it has to supply the gas. (Two years later the core of Halley's comet would finally be observed by the Soviet Vega and European Giotto satellites.) Comets eventually burn out, especially if they go close to the sun. Their supply of water and methane gets used up.

A comet is not completely water, ammonia, and methane. Mixed in with these frozen gases are dust and rock. Fred Whipple, the great scholar of comets, has convinced most scientists that about a third of the comet material is made up of rock and minerals, which don't vaporize easily. He calls it a "dirty snowball." This is a particularly vivid image for someone

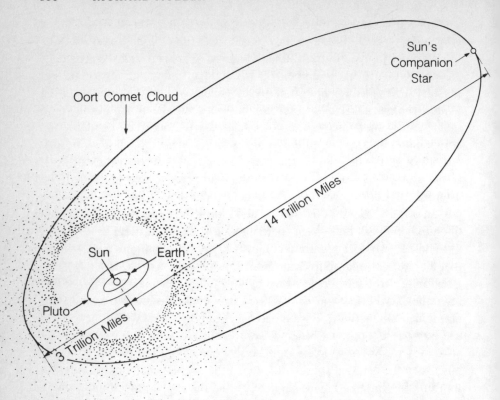

XBL 842-9423

The new picture of the solar system, including the companion star. The inner part of the Oort comet cloud is the empty "eye" of the storm. The size of the sun and the orbits of the planets have been greatly exaggerated.

like me who grew up in the South Bronx, where every snowball was dirty and where some of the tougher kids put rocks in their snowballs, just to give them a little more sting.

Of course, comets have not always been understood. They came so rarely that in ancient times their appearance had been met with fear and anticipation. It had always seemed plausible to me that the Star of Bethlehem the Three Wise Men chose to follow had been a comet. They must have known it was a comet and not a star, but it was probably the first one any of them had seen, and they ascribed great import to it. The Bayeux tapestry, portraying William's conquest of Britain, shows a comet (probably Halley's) appearing before the invasion. Edmund Halley had been the first to realize that some comets return on a regular schedule, and his comet has become the most famous of them all. It had last come close to the sun in 1910, the year my father was born. I had learned, as a child, that it would next return in 1985, and it did.

There was a lot I didn't know about comets, and Piet explained some of it to me. The great cloud of comets in the outer solar system had been named after its discoverer, Jan Oort, a Dutchman, like Piet. Oort demonstrated that the comets were not falling in from infinity, but from a region of space between $1/2$ and 1 light-year away. There had to be at least 100 billion comets out there, a virtual storm of comets. Their orbits, unlike the asteroid orbits, are not circular, but almost random. Some stay far away and are never seen, their typical 1- to 10-mile diameter reflecting too little sunlight to make them visible from the Earth. A tiny fraction of them have orbits that bring them close to the sun, but these don't survive very long. Some are burned up, and others come close enough to the orbit of Jupiter or Saturn, the heaviest planets in the solar system, to receive a gravitational kick that sends them forever away. So our nearby part of the solar system, the inner solar system, doesn't have its fair share of comets. The sun, Jupiter, and Saturn keep sweeping this region clean. We are in the "eye" of the comet storm, and, just like the eye of a hurricane, it is quiet.

Why do we see a few comets now and then? And what about Halley's comet, which comes back every 75 years? Halley's comet is an oddity, an extremely unusual comet. It had by accident been deflected into an orbit that was relatively immune to the heat of the sun and to perturbations caused by the large planets. Such orbits are rare, and that is what makes Halley's comet special. It never gets too far away, only about as far as Pluto, and so it can come back twice within one human lifetime. (One friend of mine, Fred Crawford, saw Halley's comet for the second time in early 1986.) Halley never gets close enough to the sun to burn up completely or be torn apart by the tidal forces of the sun's gravitational field.

Oort realized that most of the comets discovered by astronomers are kicked into the inner solar system by passing stars. Of the 100 billion comets that have their orbits perturbed, only a few will be aimed toward the Earth's orbit. Most comets are too dim to be seen except through telescopes. Every ten years or so a comet big enough to be seen with the naked eye passes by the Earth, and these become the comets of the seers and the newspaper science pages. A few of them get captured into small orbits, and come back a few times. And once every few thousand years a comet gets trapped into an unusually stable orbit, and comes back several dozen times, like Halley's.

If the sun has a companion star that comes into the Oort comet cloud every 26 million years, then it could kick additional comets toward the sun. Could it kick enough so that one would likely hit the Earth? This was the question that now excited Piet, Marc, and me. Piet began by saying that the number of comets in the Oort comet cloud was highly uncertain. Oort himself calculated the number at 100 billion, but those were only the comets far enough out to be affected by passing stars. If we included the comets with medium-sized orbits, 10,000 to 20,000 astronomical units in radius, then there could be 10 trillion. All of these comets would be perturbed by an orbiting companion star. Piet reminded us that these big numbers were based on the dozen or so comets that astronomers discovered every year, and any extrapolation from a dozen to 10 trillion was bound to be uncertain. But we decided to take the 10 trillion number (10 followed by 12 additional zeros) as the best starting guess. We knew that we could assume a different value, if necessary to make the theory work, since nobody *really* knew the right number.

It was Piet's turn at the blackboard. He said, "Let's take a typical orbit for the companion star, eccentricity 0.7, nothing unusual. Half of all random orbits should have an eccentricity this big. Besides, high eccentricity really doesn't help our theory, anyway, since the Oort cloud is so large." He quickly calculated that with an eccentricity of 0.7, making the orbit have roughly the same shape as an egg, the star would come only within about one half light-year of the sun, about 30,000 times as far away as the Earth is from the sun. Even at its closest approach, it would not look to earthlings like a second sun, only like another star. He estimated that this was close enough to perturb the orbits of virtually all the comets by a slight amount. How many would enter the inner solar system? The orbit of the Earth is tiny on this scale, and only a tiny fraction of the comets would come our way. Piet calculated, with a little help from Marc and me, that only 1 in 10,000 would come in. But there were 10 trillion comets, so 1 in 10,000 still meant a billion comets. A billion comets would enter the orbit of the Earth over a period of perhaps a million years. That would mean 1,000 per year, 1 new comet bright enough to be seen every three days. Not literally as numerous as the water droplets in a rainstorm, but very numerous for comets. The night sky would be spectacular. It is doubtful that any human has ever seen such a spectacle. Perhaps the dinosaurs had.

Is a billion comets enough? We quickly calculated the area of the Earth's orbit and the area of the Earth. The ratio was a half-billion. So with a full billion comets entering the orbit, we could expect 2 of them to hit the Earth. Sometimes it would be 3 hits, sometimes 2, sometimes 1, sometimes 0. On the average, 2 would hit, causing tremendous devastation and trauma. This was it. The numbers worked out. We didn't even have to fudge the number of comets. The theory simply worked!

It was still possible that we had made some mistake, even though there had been three of us checking everything. It was much more likely that we had overlooked something. Perhaps somewhere in the vast universe of astronomical data there was some fact, some measurement, that showed our theory to be wrong. We had hypothesized that there is another star orbiting the sun. Could we be certain that astronomers could have missed such an important object?

The three of us were sufficiently familiar with astronomical measurements to feel quite sure that even such an "obvious" thing as a star orbiting the sun could have been missed. Marc pointed out that the most common type of star in the Milky Way is the type called a "red dwarf." Seven out of 10 of the nearest stars are red dwarfs. Proxima Centauri is a red dwarf. Although it is the closest known star to the sun, it is, by a factor of 100, too dim to be seen by the unaided eye. It is orbiting the bright star Alpha Centauri just as we supposed our star is orbiting the sun. Proxima and Alpha are 100,000 times as far apart as the Earth is from the sun.

Why not suppose that the companion star is a red dwarf? Right now the star would be about as far away as it ever gets, since it is almost half a period since the last mass extinction on Raup and Sepkoski's plot, 11 million years ago. There are nearly a half-million more distant stars that would appear brighter than the companion, stars that are farther away but intrinsically hotter and brighter. Very few of these stars have had their distance from the sun measured. A nearby companion star could easily have been lost in the confusion.

I suggested that we write a paper right away. The theory we had found was so simple and obvious, once we found it, that I was sure somebody else would come up with the same idea. This was especially true now that the periodic-extinction work of Raup and Sepkoski had been publicized in the Sunday newspapers. Instead of the several months it usually takes to write a paper, we should give ourselves one week. Piet said that he was leaving the next morning, although he could continue working on the paper in Princeton. Marc said he was leaving in one week for a half-year sabbatical at Santa Barbara. So the one-week schedule just barely fit into his, assuming that he had nothing else to do before he left.

Since Piet was leaving town, I suggested that he write the first draft that afternoon. That way we would have all his thoughts on paper. I knew how lonely it is to be away from home, so I invited him to have dinner at my house that evening.

I promised Marc that I would deliver a copy of Piet's draft to him as soon as I had made my additions and changes. Marc took on the job of searching the literature to try to find evidence that could prove our theory wrong. He would play the devil's advocate and try to see if some astronomer somewhere had sometime done an experiment that ruled out the possibility that the sun had a companion star. I took on the job of trying to find predictions from our theory that could be used to test it. We broke up our small meeting.

As I left I had the strange feeling of not knowing whether I had just participated in something historic or had just wasted more time on another theory that would be disproved in a few days. The fact that experienced astronomers like Piet and Marc were participants made me think that this time we were really on to something. It couldn't be too obviously wrong.

Still, I worried about my plan of sending off our paper within a week. That meant we wouldn't spend as much time as usual going over the details, discussing them with our colleagues and with other scientists, searching for the tiny mistake that might make the whole paper look foolish. We would just have to work extra hard that week, Christmas week, not a good week for working hard. If we could make sure there were no really stupid mistakes in the paper, that would be good enough. A wrong theoretical paper, if clever enough, doesn't hurt your standing among your colleagues. So if we were careful enough, we would have little to lose.

The realization that it was all right to be wrong put me somewhat at ease. Yet the thought that Marc, Piet, and I were the only three people in the world who knew about the Earth's companion star and the role it had played in evolution was overwhelming.

I was anxious to try the new theory on someone else. I drove up the hill to the laboratory, where I finally found Luie having lunch in the cafeteria. I waited a few minutes to see if he would ask me again if I had made any progress. When he didn't, I volunteered that I had met with Marc Davis and Piet Hut that morning, and that we thought we had a solution to the periodic-extinction problem. I explained that it really was very simple. A companion star perturbs orbits in the comet cloud, filling the normally comet-empty region of space near the Earth every time the companion star passed near the sun. Enough comets rain down in the vicinity of the Earth so that one or more is likely to hit.

Luie didn't seem to react at all. He asked me to explain what I meant by the "normally comet-empty" region of space. I went into a long explanation of the distribution of comets and of how Jupiter and Saturn keep the inner solar system clean. Then he asked me about the stability of the orbit. I explained to him that *this* orbit should be stable. Several graduate students at the table seemed to be lost. As I went into more and more detail, I began to realize that our theory wasn't as easy to understand as I had thought. I had been working on it so hard that many of the difficult points had begun to look simple to me.

Nobody at lunch seemed very impressed, including Luie. I had the feeling he was treating me politely, as if he didn't want to interfere with my enthusiasm. I had to admit to myself that in explaining it to everyone it had not come across as being a simple theory. They thought it looked somewhat contrived, difficult, maybe obscure. I realized that they had not previously known about Oort's theory of comets, and they probably assumed that the comet cloud was an additional speculation of our theory. They thought we were making up the part about Jupiter's and Saturn's keeping comets away from the Earth, and about passing stars being responsible for the new comets we see every year. It was a large mouthful to try to take in all at once. When we write the paper, I decided, we had better make it absolutely clear how minimal our assumptions were, how most of what we were saying was established, known by the comet experts to be fact. Still, I was a little disappointed by what seemed to be almost no reaction from Luie.

That evening, over dinner, Piet told Rosemary and me about a paper he had written with the astrophysicist Martin Rees. They had analyzed the possibility that one of the large new particle accelerators being built to study subnuclear particles could accidentally trigger a "phase transition" of the vacuum of space. According to modern particle theory, the vacuum is not necessarily stable, but can be triggered to change its state, much like water does when it turns from liquid to ice. This sudden freezing would sweep over all space with the speed of light, destroying in an instant not only civilization and life, but even matter itself. Nothing within the laws of physics ruled out this possibility. Happily, Hut and Rees had calculated that such an event was extremely improbable.

Piet told us that he had carefully considered the consequences of publishing a paper like that. He had worried that politicians might use it as an excuse to delay the building of accelerators, and that colleagues might use it as evidence that Piet's research was frivolous. He could afford one such paper in his career, he decided. But now, a second paper, speculating on a companion star to the sun? That would make two "nut" papers in a row. He wasn't sure his career could stand it. I wasn't sure he was kidding.

Piet showed me a paper he had found in the library by J. G. Hills, an astronomer at Los Alamos. I read it with fascination. Hills had looked in some detail at what happens when a star passes close to the sun. He showed that the Oort comet cloud would be disturbed, leading to what he called a "comet shower." It predated our work by several years, and Hills had worked out the numbers in great detail. This paper made it possible to simplify our paper by referring to Hills for all the details on the effects of the star on the comets. Hills had not considered the possibility of a companion star, only of random passing stars, so the mathematics had to be modified a little. But to my surprise, as I read to the end of his paper, I found that he mentioned the Alvarez work on the Cretaceous-Tertiary boundary, and even speculated that a passing star could have triggered a comet shower that led to the death of the dinosaurs. What Hills had missed, because Raup and Sepkoski's piece of the puzzle had not yet been put in place, was that the showers would be "periodic," and the passing star a "companion."

After Christmas I met with Marc, and he told me what he had found about previous searches for nearby stars. As we had suspected, there had been no complete survey of the distances to stars as dim as the companion could be. That meant that a star could be orbiting the sun and astronomers never would have noticed. Later we found that the real experts on nearby stars had been well aware of this possibility. Peter van de Kamp, who had made measurements indicating that the nearby star known as Barnard's star had a planet, was one such expert. I found a review article by him in which he explicitly stated that it was possible that a small red star could be orbiting the sun without our knowing it.

Marc had also talked to his friends and colleagues in the Astronomy Department, his own collection of devil's advocates. Several of them were skeptical about the stability of the large orbit, almost 3 light-years across, but they couldn't show we were wrong. More than one such skeptic said to me, "It is surprisingly difficult to show that you are wrong, Rich." By that they meant that they had failed to find a flaw, but weren't yet "convinced."

Marc and I took turns working on the paper. I managed to find several predictions that could be added. The duration of the comet storm would be 1 or 2 million years, the time it took the star to pass by the sun. In this time the Earth would be hit by several comets, but the number could fluctuate wildly. During some storms it might be hit by only 2, 1, or even none. During others it might be hit by 5 or more. If we were to look carefully at the extinction record in the rock, we wouldn't always find just one iridium level, but should sometimes find several, one for each impact. This was the most important prediction.

I also thought about craters, particularly those on the moon that I had spent so much time studying through my 6-inch telescope as a child. Certainly the comet storms should leave scars on the surface of the moon. I mistakenly thought that the effects of erosion made it impossible to find impact craters on the Earth, except for the very young Meteor Crater in Arizona. If we could date the moon's craters, we should see a cycle similar to the one seen in the extinction data: 26 to 30 million years. I thought about adding this prediction to the paper, and decided against it. We probably wouldn't have a new moon expedition for several more decades, so we would be making a prediction that could not be easily verified or disproven. I thought it would cheapen the paper to put in predictions that could not be tested in the near future.

I was excited, very excited, but it was more of a draining sensation than a pleasurable one. The excitement was like the feeling you get from drinking too much coffee, or like the sugar high you get after eating an entire box of chocolate chip cookies. Excitement is fun when it lasts just a few minutes. The duration of a roller-coaster ride is just right, especially since you "know" you are safe. Extended excitement, especially when accompanied by doubt, is uncomfortable.

While working on the paper I had an inspiration that gave some temporary relief from the anxiety and sleeplessness the new theory had brought. The new star was more important to us than any other in the sky. It had been a driving force in our own evolution. It had eliminated the dinosaurs, the animals that had successfully suppressed the mammals for more than 100 million years. It would be coming back, to try to eliminate *us*. Didn't the scientists proposing such a star have the prerogative to suggest a name for the star? I had always admired Murray Gell-Mann's naming of the subnuclear particle, the quark, based on a line in James Joyce's *Finnegans Wake*: "Three quarks for Muster Mark." I didn't know what the line meant, although the "three" was clearly appropriate, since there were supposed to be three subnuclear particles in the proton. Obviously Gell-Mann was a scholar. Now was my chance. I told Rosemary about this unique opportunity, and said that we must find a scholarly name. She just smiled. No name immediately occurred to me. I thought back to my education in the classics at Columbia, but couldn't remember anything appropriate. It had been nearly twenty years since I had read an English translation of the *Iliad*. Nothing jumped to my mind.

I pulled out some old books from my undergraduate days and started thumbing through them. I had underlined many passages, but I found nothing relevant. I picked up a book on religion and looked in the index under death and destruction, Day of Judgment, creation, and everything else I could think of. I began to find cross-references. I read through the entire glossary of *Bulfinch's Mythology*, thinking that I should never admit that my classical education was so weak that I had to resort to this method. I rejected the name Shiva, although it was obviously appropriate. It was both the name of the Hindu god of life and death and the name of a period of mourning in Judaism. But I couldn't use it because it was already in use in the world of physics as the name of a multiarmed (like Shiva) high-energy laser at the Lawrence Livermore National Laboratory. Then the name popped out at me, Nemesis, the Greek goddess whose job it was to make sure that no earthly creature (e.g., the dinosaurs) ever challenged the dominance of the gods. To me the name reflected the fact that the dinosaurs were successful creatures who were destroyed by an event from the heavens. Rosemary liked it, so I added a footnote to our paper suggesting the name. But I was intrigued, and I kept looking. I found two other appropriate names, Kali and Indra. I added them, although I expected my coauthors to delete the footnote as soon as they read it.

The next day I showed Marc the footnote. He liked it and, to my surprise and delight, did not suggest deleting it. Piet also liked it. It read:

If and when the companion is found, we suggest it be named NEMESIS, after the Greek goddess who relentlessly persecutes the excessively rich, proud, and powerful. Alternative names are: KALI, the "black," after the Hindu goddess of death and destruction, who nonetheless is infinitely generous and kind to those she loves; INDRA, after the vedic god of storms and war, who uses a thunderbolt (comet?) to slay a serpent (dinosaur?), thereby releasing life-giving waters from the mountains; and finally GEORGE, after the saint who slew the dragon. We worry that if the companion is not found, this paper will be our nemesis.

I had added the name George and the final sentence lest the reader think we were taking ourselves too seriously. I hoped the tongue-in-cheek humor of the footnote would preempt criticism that we were naming an undiscovered object. Actually there is a history of naming objects of search in physics before they are found, such as the Greek "atom" and, more recently, Gell-Mann's quark. Nemesis was clearly our favorite of the several possibilities, and soon afterward we found ourselves referring to the hypothesized companion star to the sun by this shorter name.

My self-imposed deadline of one week was fast approaching. I had talked to Luie several times in the past few days, and he finally realized that I had a serious model to explain the periodicity. But he had never checked my numbers or ideas with any care. When I felt the paper was in good enough shape, I wanted to give it to him for a detailed critique. I expected him to be tougher on it than anyone else, so I had saved him for last. One evening I called him at home and said the paper was ready for him to bloody up with red ink. He suggested I bring it right over.

A half-hour later I rang the doorbell at Luie's large house in the hills on the north side of the campus. To my surprise, I found Walt waiting in the living room. Luie said that he thought Walt would be interested, so he had invited him over. I wondered what Luie had told Walt to get him to come over so quickly and with no advance notice. I sat on Luie's sofa, watching, as Luie and Walt read the paper.

Luie had a few suggestions on grammar and style, and pointed out two sections that weren't particularly clear. Other than that, all he said was that the paper was "interesting." I sighed in relief. Luie was usually the toughest critic I could find, and he had found no fatal flaw.

Walt said nothing about the correctness of the theory, or its importance, but, instead, immediately started looking at the consequences for geology. "Rich, do you know about the multiple microtektite levels at the Eocene-Oligocene boundary?" he asked. I didn't. In fact, at that time I couldn't even pronounce Eocene-Oligocene, although I knew it was the time of one of the mass extinctions, about 35 to 39 million years ago.

As I mentioned earlier, tektites are bits of glass found strewn on the ground or incorporated in rock. Most geologists now believe they were created when a large meteor or other extraterrestrial object hit the Earth. The impact melted sand and rock and threw it high into the sky. As the melted material fell back to Earth, it cooled and formed into glass. Microtektites were simply microscopic versions of full-sized tektites, but they were much more abundant, and could be found in sedimentary rock. Several separate layers of microtektites had been found at the Eocene-Oligocene boundary by paleontologist Gerta Keller, Walt told me. The existence of multiple layers had been a puzzle, since the layers were separated in time by hundreds of thousands of years, and could not have come from a single impact. Multiple impacts were not supposed to occur that close to each other in time, so the microtektite layers had not been interpreted as favoring an impact theory. Walt had correctly realized that our theory offered a simple, elegant explanation.

It was almost as if a prediction of our theory had come true. Unfortunately, it wasn't exactly the same, because I had not *predicted* multiple microtektite layers. I decided to mail our paper the next day, for publication in the journal *Nature*. I could look up the papers on the microtektites afterward, and perhaps add something to the paper when it came back from the referees. The most urgent thing, now, was to get as early a submission date as possible. On December 30, 1983, one day late according to my original self-imposed seven-day schedule, I went to the central post office at Berkeley and sent the paper by Express Mail, guaranteed one-day delivery, to the offices of *Nature*, in Washington, D.C. It arrived five days later.

Later that day, Walt telephoned and asked if I could stop by his office. He had something to show me, something potentially exciting. I told him I would try to make it, but by the time I broke free from my routine tasks, it was time to pick up Betsy and Melinda from their preschool. The next day was Saturday, family day. Sunday was New Year's Eve. I forgot about Walt's call. I had just mailed off the most important paper of my life.

11. Craters

ON MONDAY, Walt telephoned, and I suddenly remembered that I was supposed to stop by his office the previous Friday. Walt had been sitting on something exciting all weekend. He told me that maybe I was lucky I hadn't made the visit, since it might have ruined the tranquillity of my New Year's weekend, just as it had ruined his. He was right.

Walt had realized that if our Nemesis theory was correct, we should be able to find evidence in the record of impact craters on the Earth. He had looked at the dates of known craters, and he thought he saw some intriguing hints. Luie was skeptical, and Walt wanted to discuss the "hints" with me.

I knew that there were impact craters on the moon, but I thought that there was only one on the Earth, Meteor Crater in Arizona. Walt explained that I was quite mistaken. There were eighty-eight large craters on the Earth that had been definitely identified as having been caused by impacts of large extraterrestrial objects. If I had read the entire conference proceedings of the 1982 Snowbird Conference, I would have seen a paper by Richard Grieve in which the size, location, and ages of these craters had all been compiled. Walt told me that Meteor Crater was just a tiny pockmark compared with the many larger holes found all over the world. The craters ranged in size from a few miles to over 100 miles across. I had never learned of them mostly because none of the large ones

XBL 875-2054

Areal distribution of known impact structures. Open symbols represent craters with associated meteorite fragments. Closed symbols represent structures with shock metamorphic effects and in some cases siderophile anomalies.

was 65 million years old, so none of them could be the impact from the Cretaceous catastrophe. Until recently, the Cretaceous event had been the Alvarez group's sole interest.

Walt hadn't taken the periodic catastrophes any more seriously than had Luie, until we had written our Nemesis paper. It is difficult to accept observations that make no sense. Like almost everybody else, he thought that the "effect" of Raup and Sepkoski was probably a spurious result of a statistical fluctuation. At least, that was how he felt until he read how our theory could make sense of it all.

There was another good reason why Walt was now ready to join those of us who accepted the periodic mass extinctions. Several years earlier, geologists Alfred Fischer and Michael Arthur had written a paper claiming there was something strange happening on the Earth every 30 million years. Raup and Sepkoski had properly referenced Fischer and Arthur's work in their own paper, and I had even looked it up and tried to read it. I didn't have much luck in understanding it because it was written for geologists, not for physicists, and I didn't know half the technical words.

So I had ignored this paper. But Fischer had been one of Walt's geology professors at Princeton, and more recently had been part of Walt's paleo-magnetism-paleontology collaboration in Gubbio, Italy. Walt understood the Fischer-Arthur work in detail. But he had never taken it seriously, in part because the effect was very weak, from a statistical point of view. And, besides, it didn't make any sense.

Walt had been quick to recover from his previous, perhaps too casual dismissal of this paper. Almost immediately after reading our Nemesis paper he realized how our new theory fit in with the earlier work. Now it made sense. He soon extended his understanding to make new predictions, which could be tested.

No one had found the crater associated with the Cretaceous impact. The Nemesis theory indicated that most of the other catastrophes would have craters, too. Moreover, there wouldn't be just a single crater from each crisis, but several, since the comets came in showers or storms. For any given crisis we should be able to find at least one of the several craters. Walt had looked at the list of craters in Grieve's compilation and compared the dates of the impacts with the dates of extinctions. There were some problems, but also some hints that made him think more detailed analysis was worthwhile. He wanted to know if I was interested in looking at them with him. I certainly was!

As I walked into the Earth Sciences Building the next day, I noticed the display of fossils on the ground floor. Most striking was a complete skeleton, still embedded in rock, of a Parasaurolophus, a large plant-eating dinosaur, hung on the wall. In a glass case was the skull of a Tyrannosaurus. They were the first dinosaur fossils I had seen since I had become seriously involved in the question of their extinction. Now these fossilized bones had a personal meaning to me. The death of the dinosaurs had become, for the first time since my childhood, something important in my life. I felt a new relationship to those bones in the case and on the wall.

Walt was waiting for me in his office, and he immediately offered me a cup of espresso. His office had two parts, a back room with his desk, and a front room filled with files and cabinets and charts and boxes. There were several physics books mixed in with his collection of geology texts. The rooms had the clutter typical of a professor, but in addition to the piles of papers and reprints there were dozens of rocks of various sizes, on shelves, in boxes, on tables, and on the floor. I asked Walt about them, and he pointed to several piles in sequence and described the project associated with each. One pile of rocks had been collected in an attempt to understand microplate tectonics in the Mediterranean; it was waiting to be analyzed. Another pile of rocks, each one in its own plastic Ziploc bag, had been collected during the previous summer near the Cretaceous boundary layer. Walt was trying to understand the conditions just before and after the catastrophe. Looking at the pile of rocks he had lugged back from Europe, I decided I would never complain again about the weight of the scientific papers I carried home from my European trips.

Walt said, "Okay, let me show you what I have." He pulled out the large, red-bound *Snowbird Conference Proceedings*. In it was a map of the eighty-eight known impact craters on the Earth. "You'll notice that most of them are in Europe and North America. Can you guess why?" I thought of rotations of the Earth, but had no luck in figuring out why impacts would be more likely in the Northern Hemisphere. Walt didn't let me waste time for long. He said, "Because that is where the most geologists live." Impact craters don't go unnoticed if they are in a geologist's backyard—unless they are too large to notice, that is. Walt showed me a photo of the Manicouagan crater in Quebec, taken from a satellite. All that could be seen was a water-filled ring, 80 kilometers across. This huge crater was 210 million years old. The ring structure was first noticed in 1975 in satellite photos, but only after a dam filled it with water. This reminded me of a *MAD* parody of King Kong, in which a ship's crew is standing in the middle of a huge footprint of Kong, but is unable to see it because it is so large.

Walt pulled over several large sheets of graph paper from a corner of a big table in the center of his office. "The horizontal scale is time, in millions of years," he explained, "with the present here at the right-hand edge, and 250 million years ago at the left-hand edge." On this scale he had placed a cross to represent each crater. If there was any strong 26-to-30-million-year periodicity in the plot, it certainly wasn't immediately evident. I was mildly disappointed.

"What I realized, Rich, was that many of these craters are very poorly dated. Look at this." He had marked a page in the *Proceedings* that was filled with a large table, which I soon saw was a list of "impact craters" on the Earth. He showed me that the uncertainty in most of the crater ages was 50 million years or more. That had been good enough to determine the rate at which large objects had been hitting the Earth over the past billion years. But if some crater ages are off by 50 million years, they can totally wash out a 26-million-year periodicity.

Walt's simple solution to the problem was to ignore all craters that didn't have accurate dates. As long as these craters were not chosen on the basis of whether or not they fit a certain periodicity, there could be no harm in this procedure, no systematic bias introduced. Walt had selected the accurately dated craters, of which there were about two dozen, and had marked their ages on his plot. He pointed to a clump of craters that had all been gouged out about 37 million years ago. "That's at one of Raup and Sepkoski's mass-extinction events," Walt said, "the Eocene-Oligocene. What's more, Popagai, the biggest crater on my list, falls right there." Popagai is named for the region in Siberia where it is located.

We started paying more attention to the larger craters on the list. The next two largest, Puchezh-Katunki, in Russia, and Manicouagan, had ages of 183 million years and 210 million years, respectively, with uncertainties of about 5 million years. They missed the expected dates of Raup and Sepkoski's extinction cycle, yet the difference between their ages was 27 million years, intriguingly close to 26. The next two largest craters, at Steen River in Alberta and at Boltysh in the Ukraine, occurred approximately 95 million years ago, about 30 million years before the Cretaceous catastrophe. The next two largest, at Rochechouart in France and Gosses Bluff in Australia, didn't seem to fall at the right times, but they were spaced by 30 million years. How many coincidences like this were reasonable to expect if the crater data had nothing to do with extinctions?

Both Walt and I knew that it was easy to fool ourselves with a qualitative study such as we were making. It was tempting to note everything that agreed with periodicity and ignore anything that didn't. But we couldn't help but get excited. We could do a detailed mathematical analysis later, although we knew that it was unlikely to help much when we had only two dozen craters.

I suggested that we replot the data as a histogram, a kind of bar graph, giving more emphasis to the larger craters. Walt pulled out a clean sheet of graph paper, and I set to it. About five minutes later we had a clean new plot. There seemed to be periods of intense cratering with a spacing between them of 26 million to 30 million years. I didn't know whether to believe my eyes. A scientist's main strength is skepticism, particularly self-skepticism. I had to be my own toughest critic. But I couldn't restrain myself from saying, "Looks good to me." Walt responded, "Me, too."

We wrote down the ages at which the clumps of craters occurred, and calculated the average time between them. It turned out to be 28 million years, right in the middle of the 26–30 range that Raup and Sepkoski had found for the period of mass extinctions. It was still somewhat bothersome to me that some of the ages missed the dates of the extinction events. Walt reassured me. "The time scale of the extinctions is very poorly known," he said. "It could be off by as much as 10 million years. But the time in between extinctions can be much better determined than the absolute time."

Walt was certainly right about one thing. Had we had this meeting on the previous Friday, as he'd wished, it would have spoiled my whole New Year's weekend. I would have spent that time analyzing the craters and thinking about mass extinctions rather than enjoying my family. I couldn't have concentrated on the Rose Bowl game.

I suggested that the next step for me would be a more detailed and careful statistical analysis on the large computer at the Lawrence Berkeley Laboratory. I had gained considerable experience in the required techniques while working on projects in particle physics and astrophysics. Walt decided that the best thing for him to do was to look more carefully at the crater data. He was certain that some of the craters had had their ages more accurately dated since Grieve made his compilation. We both knew that there would be a danger in going back to the literature for better crater ages. If we let ourselves have the power to accept or reject ages, it would be nearly impossible to do it in an unbiased way. We would know which crater ages fit in well with the extinctions, and which didn't. How could we avoid the tendency to reject those ages that didn't fit our hypothesis?

There was also a danger that in being "honest" scientists we would lean over backward to accept ages that contradicted our result, even if we knew that they had been poorly determined. That could have the effect of washing out a real effect, of making it go away. Our case would really have to rest on the list compiled by Grieve, since he had done the work before any hypothesis of periodicity existed. Walt's scholarship in the library would be necessary, however, if only to avert criticism that we were being lazy. But Grieve's unbiased list would have to be the basis of our analysis.

I felt a bit stupid that I hadn't known enough to look at craters on the Earth, but I consoled myself with the thought that the person who had compiled the list, Grieve, had not noticed the periodicity. There was no way to notice it until you threw out the craters with inaccurate dates, and there was no reason to throw them out unless you were looking for something like a short-term periodicity. Sometimes you have to know what you are looking for in order to see it.

My claim to Walt that I could do a statistical analysis on the large laboratory computer was based on the assumption that I could find someone to help me run my programs. Although I was a good programmer, I no longer knew how to enter a program into the big computer. I had spent hundreds of hours working with large computers as a graduate student, and had grown to hate the experience of fighting with the big machines. As soon as I had the chance, I had stopped using the computer for anything of my own. I had hired a full-time programmer, and had several graduate students and colleagues who could do whatever computer work my research required.

I walked to the small set of interconnecting offices that the members of my research group and I occupied. To my disappointment, none of the real computer experts were there. I waited. Suddenly Saul Perlmutter walked into the room. I pounced.

"Saul," I said, "I have something very exciting, and I need your help." I quickly told him everything that Walt and I had done that morning, and Saul became enthusiastic, as I had hoped. The first task would be to get the data into the machine. I volunteered to sit and type at the computer terminal if he would set the programs up to receive the data. I sat and typed for the next forty-five minutes, while Saul went down to the computer center and explored the vast library of programs available to us. He came back just before I finished. He had found the library programs for Fourier transforms and was all set to try them out. I watched him for about a half hour, as he gradually worked out incompatibilities between the formats of the various programs. How close was he to getting it running? Saul answered that it shouldn't take more than a day or two. I was dejected and walked into my office figuring I could spend some of the time reading papers I had never read, such as Grieve's on the crater ages.

That night I had trouble falling asleep, once again. This insomnia has occurred only during the most exciting periods of my life. Some people can't sleep because they can't figure out how to pay the bills, or how to solve other personal problems. I couldn't sleep because I couldn't turn my brain off, or maybe because I didn't want to. I kept on thinking about the craters and ways of analyzing them. How could I take into account the uncertainty in the crater ages? Had Walt and I fooled ourselves? Or did the crater data really prove that we were on the right track? If so, what were the further implications? What did all this imply for the theory of evolution? For the solar system? The excitement was tiring, almost exhausting, but it still kept me awake. I decided I had better stop thinking about it, and think about something physical, not intellectual, like mountain climbing or skiing. I imagined myself making perfect parallel turns down a steep mogully slope. . . . Why not try a Gaussian ideogram of the crater data, and Fourier-analyze that? Right, that's the best thing to do. I'll help Saul set it up first thing in the morning. Should I emphasize the bigger craters? Better try it both ways. I wasn't falling asleep.

By the next day, Saul had the program running and was producing beautiful plots on the computer drafting system. We used Fourier analysis, a mathematical technique that looks for periodicity in data. The first plot showed a strong peak right at the period of 28.4 million years, high above the background levels at other periods. That was it, what we had seen. Now we had to determine whether it was statistically significant, or whether it could have arisen by chance. I explained to Saul how we could do this. The technique was called the "Monte Carlo" method, after the famous gambling city in Monaco. We would program the computer to generate ages randomly, dates that had nothing at all to do with the times of the extinctions or the real ages of the craters. We would then have the computer assign these to the real craters in place of their true ages. Next, we would have the computer redo the complete Fourier analysis, having it look for a periodicity in the ages. There should be none, since the dates were randomly chosen. After that we would have the computer pick another set of random dates and again redo the analysis. And again, and again. We would have the computer redo everything over and over until, by sheer luck, it produced a peak in the Fourier transform plot as large as the one we saw in the actual data. It would be a statistical fluctuation, but if we kept trying, it was bound to happen. It might take a long time to get a random sequence that looked periodic, but the computer wouldn't get bored. I wanted to know how frequently a statistical fluctuation would simulate a periodicity as good as the one Walt and I had found.

It took the computer less than half an hour to produce a run with random ages that showed a periodicity as good as the one we had found, and a half hour later it randomly generated a set of dates even *more* periodic than ours. That showed that it was possible for our effect to be spurious, something we always knew. But it took from 100 to 500 tries to get one as good as the real data. Put another way, that meant that the odds that our apparent periodicity came from a random sequence were between

100 to 1 and 500 to 1. Pretty good odds, I thought. The effect was real. The cursory analysis that Walt and I had made the first morning was vindicated.

A few minutes later, Luie walked into the office, wondering if we were getting anywhere. I thought it was excellent timing, just after we had the first stage of the computer analysis complete. I showed him the crater plots, the Monte Carlo analysis, and the Fourier transforms.

Luie was unimpressed. Odds of 100 to 1 were not very good, not for an important discovery. How many ways were there to analyze the data, he asked. If you could look at the same data one hundred different ways, then the odds were that at least one of them would show an effect that had odds of 100 to 1. I protested that I knew that already, and had taken that danger into account in the analysis. We had accepted random peaks over a wide range of periods, not just those near the 26-million-year one we had hoped to find. But Luie didn't back down; 100-to-1 odds, he said, are the same as "two and a half standard deviations," in the jargon of statisticians. A good physicist doesn't even pay attention to anything less than 3.5 standard deviations. He knew of a lot of 2.5-standard-deviation effects that had turned out to be totally false when checked by repetition of the experiment.

I didn't back down either, even under Luie's strong onslaught. I argued that the 100-to-1 odds didn't even take into account the fact that the impacts making the craters took place at the same *time* as the mass extinctions of Raup and Sepkoski, at least for the accurately measured craters from the recent past. Certainly that was a "coincidence" that he couldn't ignore.

Quite the contrary, Luie replied. The dominant period in the analysis of Raup and Sepkoski was 26 million years. The Fourier transforms had shown that ours was 28.4 million years. They disagreed by 2.4 million years. Not much, but the disagreement would build up as we went farther back in time. After two cycles, the time of the extinctions would miss the time of the cratering by 4.8 million years, and after three cycles by 7.2 million years, and so on. After only 140 million years, the two cycles would be out of phase by 13 million years, a full half-cycle. The impacts must be arriving exactly halfway between the mass extinctions. Luie got a little sharper and more personal. He directed his attack right at me. "If you publish this, Rich, they will just laugh at you. It's nonsense." That comment effectively ended the conversation, at the time.

I wanted Luie's reasoning to be wrong, but I was bothered that I didn't have ready answers to his criticisms. Perhaps I wasn't being as tough on myself as I thought I was. The most disturbing point he'd made was that the two cycles, mass extinctions and craters, would fall out of phase with each other. The uncertainty in the periods was large enough for me to figure there must be some good answer to Luie's analysis, but I didn't have it at hand.

Walt came up to the lab for lunch. He had already found journal articles with improved ages for two of the craters. I told him about Luie's criticism that the crater period and the mass-extinction period would get out of synchronization with each other. Walt said that Luie had raised the same objection to him a day earlier. "Let me show you the answer I've come up with," Walt said. He pulled out another piece of graph paper. On it he had plotted both the crater cycles and the extinction cycles. Sure enough, they got out of phase at about 140 million years ago. But Walt pointed out that this was the period when the extinction data were extremely weak and the uncertainties in the time scale very large. It was likely that Raup and Sepkoski had simply misidentified two of the peaks in this region. By the time the data became accurate again, the two cycles were back in phase.

I had been making a simple mistake in the way I had been thinking about the periodic extinctions. Fourier analysis looks for *exact* periodicity, with constant amplitude. If the real data have a period that changes slowly, or have a varying amplitude, then Fourier analysis will not fit the data with a single period. In their analysis, Raup and Sepkoski had shown that two periods fit their data almost equally well, either 26 million years or 30 million years. Thus we never should have expected a single fixed period. In our Nemesis paper we had stated that the period of Nemesis should vary by about 10% over the last 250 million years because of the perturbations of passing stars; this was consistent with a superposition of the two distinct periods seen. What Walt showed in his plot was that the times of the episodes of cratering agreed with the times of the mass extinctions, even though the best fit periods for the two sets of data were not exactly the same.

Luie came back a little later and, before Walt could show him the plot, handed us copies of two papers written twenty years earlier, one announcing the discovery of the "kappa meson," and the second demonstrating its confirmation. "The kappa doesn't exist," Luie said. "This was the only particle 'discovered' by my group that ever turned out to be completely wrong." I noticed that Luie was *not* one of the authors. I assumed that he had been skeptical at the time of publication, too. I knew a little of the history of the claims of the kappa meson, but I hadn't realized that it was Luie's group that had incorrectly "confirmed" its existence. The kappa was too short-lived to be directly observable, but its fleeting existence had been deduced by bumps seen in the statistical distributions of the particles' decays. Similar peaks had been seen in two different reactions, and they gave the same value for the mass of the kappa. In one paper the peak had been high enough for the odds of its coming from a statistical fluctuation to be only 1 in 400; in the other paper the odds were 1 in 100,000. So everyone had believed in it, but it was never seen again. Perhaps one of those rare statistical fluctuations had actually happened.

"Look at the peak in this plot," Luie said, pointing to one of the papers. I looked. It was pretty big, especially for a known statistical fluctuation. "That's more impressive-looking than your peak, don't you agree?" I smiled at Luie. He was making a pretty strong point, but I didn't agree. Most true discoveries are made with odds less than 1 in 100; if he insisted that odds always beat 1 in 100,000 his group never would have made all the discoveries that led to his Nobel Prize. He was being too skeptical, I thought.

Instead of addressing his barb directly, I said, "Walt has a good answer to your question about the periods getting out of phase." Walt showed his plot to Luie and explained how the weak data and large uncertainties in the middle range could reconcile everything. Luie was impressed, and he admitted that Walt had satisfied him, at least on that aspect. "I think you should post these papers somewhere," Luie said, picking up his kappa meson papers and looking around. He finally settled on the door that connected the two main offices of our small complex, and he taped them at eye level. We left them there for several months. It was a helpful reminder that even smart people can fool themselves.

Luie had been helpful, though somewhat annoying. It was hard to accept such severe criticism, but I appreciated Luie's speed in backing down when we found a really convincing argument. With the boundless enthusiasm that Walt and I were showing, Luie's presence and his force were essential.

Although Walt had answered his biggest objection, Luie still didn't believe that our evidence for periodic cratering was statistically signifi-cant. Obviously the data didn't speak for themselves.

Or did they? Maybe I could let the data convince Luie if I could overcome his prejudice that it was all nonsense. I felt that I knew Luie pretty well. For nineteen years I had been consciously studying the way he thinks. Just what would it take to convince him? Suddenly I realized I had the answer. I quickly devised a plan that I thought would work. If this didn't convince Luie that the craters were periodic, then it could only be because they *weren't* periodic. I found Saul and described to him what we would have to do. Saul was delighted. It sounded like fun. We would trick Luie, only temporarily, but in the process Luie would convince himself that our analysis was right. We would use a trick that Luie himself had invented over twenty years earlier and that had been developed by his colleague Gerald Lynch. But we would introduce a slight variation. The procedure made use of "double-blind" techniques and was called, simply, "Game."

12. Double Blind

ON FRIDAY, January 6, 1984, I turned forty, the official beginning of middle age. I was no longer a "young scientist," although I knew I looked younger than I was. A physicist has an advantage when he looks young, since people confuse his experience with native intelligence. They conclude that he is smarter than he actually is. But the true importance of that day had nothing to do with my age. It was then, only two days after Walt had first shown me the crater data, that Saul and I had Game ready to try on Luie.

The first Game program had been written for a specific purpose. A physicist in Luie's group had been analyzing the vast quantities of data collected by the bubble-chamber team. He had plotted the amount of energy needed to create a short-lived elementary particle known as the rho meson. He had expected the plot to have a bell shape and the energy of the points near the center of the bell to be related to the mass of the rho by the famous equation $E = mc^2$. But the plot didn't appear to have the right shape. There was a dimplelike dip right at the top of the curve. He concluded that the bell was split in two. He showed it to many people, including Luie and Gerry Lynch, but nobody was persuaded. Everyone thought it was just a statistical fluctuation. The physicist refused to back down. He wanted to publish. That was when Luie came up with the idea for Game, and Gerry developed the idea and turned it into a computer program.

In order to show what statistical fluctuations really looked like, the computer was used to generate seventy-two sets of random data. (The computer was asked to generate one hundred, but the time ran out before that many were created.) Each set was designed to simulate what real data would look like in the absence of the dip in the curve. The physicist was then asked to look through the seventy-two plots and pick the ones that appeared to have the biggest splits in them. This would show him how large splits could arise from statistical fluctuations.

Twenty physicists around the lab agreed to try the exercise. Each picked out the two plots he felt had the biggest dips. The physicist who had done the original analysis thought that this didn't prove anything, because he felt that none of the "statistical fluctuations" in the fake plots was as profound as the real split he had found in the true rho data set. But he didn't know that Gerry had set a trap for him. Along with the simulated data, Gerry had secretly inserted the *real* data set, with its dimple on top. When he looked over the seventy-two graphs, the physicist did not recognize that the real data set was there, and he failed to include it among the sets he picked out as having the biggest split. Thirteen of the other physicists also picked fake data as having a deeper split than the real data. The only reasonable conclusion to draw was that the split was not statistically significant. The physicist had no other choice but to back down. Game had done its job. There is no one quite so valuable to a scientist as a good devil's advocate.

I had heard Luie tell this story of Game many times in the years I had spent with him. I obtained some of the details from Gerry, who showed me the paper in the *Physical Review* in which the use of the Game program was described. The experiment had been done "blind," in the sense that the physicist hadn't known what result he was supposed to find when he looked through the data. Blind experiments are used extensively in medicine, when the patient isn't told whether he is being given real medicine or a placebo. In fact, medical researchers have found that blind experiments usually aren't good enough. When giving a placebo, the doctor can subconsciously convey his expectations to the patient, who can react accordingly. Sugar pills don't cure psychosomatic illness as long as the doctor knows they are sugar pills, so medical experiments are usually done "double blind," with neither the doctor nor the patient knowing which medicine is real and which is the placebo until the effects of the dosage are recorded.

I had watched Luie recently use a blind procedure to test William

Fairbank's reported discovery of quarks, particles with a fractional charge equal to one third or two thirds of the electron charge. Luie prepared a random number, which was added to Fairbank's data in such a way that Fairbank would not be able to tell whether the charge he was measuring was fractional or not. Only when Fairbank had completed his analysis of systematic effects and uncertainties did Luie go down to Stanford to supply the random number. Luie and Fairbank were all set to celebrate, because this experiment would prove to the world that Fairbank's discovery was valid. But when he introduced the random number into the analysis, the results no longer showed charges of one third or two thirds of the electron charge. Blind experiments were certainly useful in Luie's world of discovery.

I was determined that my experiment with Luie would be done double blind. I would give him plots to look at. I would tell him that all the plots were random, but in fact one of them would contain the real data. Even I wouldn't know which was which. I knew Luie would enjoy the thought of this procedure being tried on him, and he would guess that I had inserted the real data. And I knew that he would try to turn it around, and use it to show me that *I* was wrong.

Saul prepared the plots according to my instructions, and I gave the set to Luie. He took them and studied them carefully. Walt watched as his father looked at each computer-generated histogram, and each computer-generated Fourier transform. Luie spoke out loud, telling us why he thought each plot looked interesting or not. He estimated the sizes of statistical fluctuations in both the histograms and the transforms by looking at the range of peak sizes in the various plots. It took him only about ten minutes to go through the twenty plots, and he announced that his mind was made up. He had picked out three plots that looked most likely to have periodicity.

"This one is the best," he announced, "and these two are tied for second place." But he wasn't going to let us think that we could trick him, so he quickly added, "And the second one here I recognize as the real data. I knew you would stick that one in, Rich."

I really didn't know which was which, so I asked Saul to look up the numbers in the secret table he had made. My heart was pounding while I waited for the answer. Luie thought the experiment was over, and that I had failed. After all, he had picked a random plot as being *more* periodic than the real data, and a second random plot as being just as periodic. Didn't that prove that the real data did not have a significant periodicity?

No, it didn't, not yet, not until we checked the secret table. Only Saul and I knew of the additional twist we had added to Game.

Saul read out the results. The plot that Luie had identified as the real crater data plot was, indeed, the real crater data plot. Kudos for Luie. But the plot that he had picked as most significant was *not* a random-data plot, and neither was the plot that he had picked as tied with the real data in significance. I sighed in relief. The trick had worked. Luie was confused. If they weren't random data, what were they?

I explained that of the twenty plots we had prepared, one contained the real data, just as he had discerned. But only sixteen of the remaining nineteen contained random data. For the other three we had made the computer simulate what data would look like if there were *real* periodicity. For these, Saul had given the computer eleven craters with exact 26-million-year spacing, rather than random spacing. To simulate the effect of uncertainty in the crater ages, the position of each crater was jittered a little on the plot according to the accuracy with which the crater ages could be measured. This had resulted in a plot that looked somewhat periodic, but not perfect. It showed what true periodicity should look like given the uncertainty of the dating techniques.

The three plots Luie had picked included that with the actual data and two with real periodicity. Luie hadn't picked any random data at all. In making his choices, Luie had actually shown that our data were as good as you could expect a truly periodic signal to be. In fact, he had missed the one other plot in the set that was also truly periodic, showing that his level of skepticism was so high that he could actually miss a true signal. Luie just nodded, and then he smiled. "You took Game further than it has ever been taken before. That was very clever." He kept on nodding. He seemed to like the idea that he had been tricked.

"What do you think now?" I asked.

"You've shown me I have to take a new look at it," he answered. I was disappointed at his noncommital answer, but in subsequent days I realized that Game had been the turning point. From that moment on, Luie's skepticism began to shrink.

I had to travel to Washington to attend a meeting relating to my accelerator-dating work. Since we had not yet written a paper describing the crater periodicity, I didn't tell anybody at the meeting about our discovery. It would be too easy for someone else to duplicate, perhaps to extend. The Nemesis paper had been sent off, and I felt free to talk about that, even though I realized that there must be other people as smart and knowledgeable as Walt who would read this paper and decide to look at crater ages. Some people at the meeting had heard about our Nemesis paper already. Word of an exciting (or crazy) new theory spreads fast in the relatively small community of physicists.

Back in Berkeley, I found that Luie was raising a new objection. "The peak in your Fourier transform comes from just two craters," he told me. He was referring to the craters Manicouagan and Puchezh-Katunki, each more than 40 miles across. "They are 27 million years apart, and so big that they dominate everything else." I told him I would look into it. I tried throwing first one crater and then the other out of the data. Sure enough, the significance of the periodicity went down. But you expect the significance to go down when you throw away valid data. Other physicists independently raised the same objection to the undue influence of these large craters.

We had several possible justifications for emphasizing the larger craters. Perhaps only the larger impacts caused mass extinctions. Maybe only the large craters were periodic, because they came from comets in the inner comet cloud, whereas the smaller craters came from comets in the outer part of the cloud. Nonetheless, I realized that we would be more convincing if we did not use such ad-hoc arguments. People would think I had biased the data, tried to make the periodicity stronger, by plotting the bigger craters with bigger boxes. We redid all the analysis with all craters plotted with boxes of equal size. The new Fourier transform still showed a strong peak, although its statistical significance was somewhat reduced. I also looked to see what would happen if we included only craters bigger than a certain diameter, and confirmed that most of the effect did, in fact, come from the bigger craters. But a disturbing new thing appeared in the plot, a second peak, at 21 million years. It was almost as high as the 28-million-year peak. I wasn't sure at first what to make of it.

Understanding that second peak took most of my time for the next two weeks. I knew I had to be careful not to dismiss it too readily as an artifact, since it might be real. Maybe there was something new going on, something additional to what we had just discovered. If the 21-million-year period was a statistical fluctuation, did it mean that the *other* peak, at 28 million years, was also a statistical fluctuation, that our previous analysis was wrong? Wrestling with these problems was not easy. We made many new Monte Carlo simulations before we finally understood the effect in detail, and I could convince myself (and nearly every skeptic, including Luie) what the second peak was. It turned out to be a harmonic (a spurious result that sometimes comes up in Fourier analysis) of the 28-million-year peak, with a period exactly three fourths that of the main peak. When we generated a real 28-million-year period, we would often get such a spurious second peak. We showed that it was very unlikely that the 28-million-year peak was a subharmonic of the 21-million-year peak by doing additional simulations. This implied that the 28-million-year peak was real, even though the 21-million-year peak was not. Two weeks of intense effort on this problem resulted in one short explanatory sentence in the paper that Walt and I finally wrote.

Walt had worked hard on the literature search. He had obtained a key to the Geology Library, and had stayed up as late as 4:00 A.M. trying to work his way through the difficult papers. His improved list of crater ages was ready, and we ran our Fourier-analysis programs on them. The results were virtually indistinguishable from those we had gotten before. Relieved, we were finally ready to send our paper off for publication. I volunteered to write the first draft.

I delivered a copy of the draft to Walt in his office. He looked at it, and said, "Rich, I think we should discuss the order of the names." I had put the names in alphabetical order, as I had on the Nemesis paper. I thought it was gracious of him to raise the subject. I replied, "Walt, I regard this discovery as a present from you to me. You were the one who realized it was worth looking at crater ages on the Earth. You thought you had found something interesting before I even heard of the idea. You could have had anybody you wanted to help you with the mathematical analysis. You could have done all this yourself." Walt said, "On the contrary, I regard this as a present *you* gave *me*. Without your Nemesis theory, I never would have looked at the crater data." "Fine," I replied. "Let's keep our mutual admiration society, and keep the names on the paper in alphabetical order."

On January 26 we put the paper in the 4:00 P.M. mail. It had been less than five weeks since Marc, Piet, and I had devised the Nemesis theory, and just over three weeks since Walt and I had looked at the crater data. Walt wanted to call Dave Raup and Jack Sepkoski, to tell them about these two papers. We agreed to do it the following day, and to get Luie and Saul to join in on extension telephones. Walt called Raup ahead of time to let him know that we wanted to talk to him as a group, but he didn't tell Raup what the subject was.

That night, with the paper in the mail, I had my first good night's sleep in three weeks. I was finally able to stop thinking about dinosaurs, craters, comets, and companion stars.

The four of us gathered in my office the next morning and spread out to the extension phones. Raup and Sepkoski were waiting for our call. Walt took the lead, and described to them the Nemesis paper and our new one on periodic cratering. Neither Raup nor Sepkoski seemed particularly surprised. "We've seen another paper proposing a companion star, which *Nature* wants us to referee," Raup said.

My heart sank. "What was the mechanism?" I asked. "What made the comets or asteroids hit the Earth?"

"Something about comet showers," Raup replied. "I don't really understand it." Someone else had independently invented our theory. I asked for the names of the authors. Raup said they were Daniel Whitmire and Albert Jackson.

"Does their paper have a date on it, indicating when it was received by *Nature*?" Everyone on the phones knew what was on my mind. I felt a little guilty, worrying about priority in the midst of so much interesting science. Raup asked me to wait while he looked.

He came back to the phone. "January 4," he said. It was the *same* day that our Nemesis paper had arrived at *Nature*.

"Can you send us a copy?" I asked.

"I really can't do that," Raup replied. "I was sent this paper to referee. I received it in confidence. But let me give you the authors' addresses. You can ask them for a copy directly."

We returned to science. Walt told Raup and Sepkoski in more detail about the crater periodicity. They had done quite a bit of analysis on the periodicity of the mass extinctions, so they knew all the pertinent statistical questions to ask, but they had not used Fourier-transform methods and they were very curious about the technique. I was surprised, since I thought the Fourier-transform method was much simpler and more obvious than the complex time-series analysis they had used. We had nearly an hour of discussion. We told them we would send them preprints of our papers and would stay in touch.

Immediately, I tried to call Whitmire and Jackson, and succeeded in reaching the former. I told him that I had heard about his paper from Raup, and that we had submitted a similar paper to *Nature*. We exchanged details, and discovered that our theories were almost identical. They had assumed a smaller mass for the companion than we had, and this had required that they make the orbit have a higher eccentricity than our value of 0.7. It was difficult to get all the facts straight over the phone, but Whitmire said he'd be happy to send us their paper. We promised to send them ours. He also asked if I would send a copy of the Raup and Sepkoski paper that showed the periodic mass extinctions.

"Haven't you seen it yet?" I asked incredulously.

"No," Whitmire replied. "I read about their work in the newspaper."

Al Jackson and he had worked together before, and they knew about Hill's paper on comet showers, so it all seemed to fit together for them. I remembered my comment to Marc Davis, early in December, on how the newspaper coverage would give us competition. But rather than feeling any jealousy, as I might have toward a competitor, I realized that I felt an unexpected affinity with these two scientists I had never met. They had been the only scientists besides myself that I knew of who had taken the mass-extinction periodicity seriously, and they had invented the same theory.

13. The Star of Rich's

W ITH BOTH papers in the mail, I had time to think about what we had just discovered. The work of Raup and Sepkoski had led us in a very productive direction. Periodic mass extinctions really happened. Catastrophes were not rare occurrences, but were repeated on a regular schedule. The dinosaurs must have been even more robust than we had previously thought. They had survived several impacts before the big one at the end of the Cretaceous period finally did them in. To kill them off took what was probably the most violent impact on the Earth since the evolution of vertebrates.

Of course the dinosaurs weren't nearly as successful as the cockroaches. These little creatures managed to survive every catastrophe over the last 250 million years, including the Permian-Triassic and the Cretaceous-Tertiary extinctions. They are very well adapted to fight extermination, as many city dwellers know.

It is possible that evolution might have stagnated without periodic cleaning of the slate. Darwin had postulated "survival of the fittest." But there is also "survival of the first." Seniority is, perhaps, an aspect of evolution that has been underrated, since until now we haven't known about the frequent blows from space. New species cannot flourish, and may not even survive, if all the ecological niches are already filled. Except for the catastrophes, would our world still be dominated by dinosaurs, or trilobites, or maybe even protozoa? Since I have no credentials in evolutionary theory, I am free to speculate. I have no reputation in that field to lose.

When the pictures of the planet Earth from space were obtained, many people commented on how isolated and lonely the Earth looked. People called it Island Earth, something we share and must take care of. But we are not as isolated as we once thought. Periodically a projectile from space attacks our island. We may have been misled by Darwin into thinking that our survival depends on competition with other creatures sharing this island. As Roger Mollander was to suggest in a *New York Times* essay a few months after our work, the existence of Nemesis, the "death star," may be just what our species needs to make us appreciate that the real threat to our existence may not be other humans.

About a week after we first telephoned him, Dave Raup called back. He had just been asked to referee another paper for *Nature*, this one by Michael Rampino and Richard Stothers, of the Goddard Space Flight Center, in New York City. They had an alternative theory, one that did not require a companion star. In their theory the comet showers were triggered by oscillations of the sun up and down in the galactic plane, which we already knew had approximately the right period. They had also found the periodicity in the craters, just as Walt and I had. Dave again felt that he could not send me his copy of their paper, but he was sure I would want to call them.

I told Dave that Marc Davis and I had considered galactic oscillations as a possible trigger for the extinctions, but had rejected them because the phase of the oscillations did not match the phase of their extinctions. Dave said that this paper showed a very good match between the two.

What had I done wrong? I looked for my notes on galactic oscillations, and finally found them buried under an enormous pile of papers on my desk. Then I looked over my calculations of the frequency of galactic oscillations and the data that said we were passing through the galactic plane right now. I struggled to find my mistake. I found nothing wrong. If there was a mistake, it must be a very deep, fundamental one, not the kind you can find simply by checking numbers.

I told Luie about the paper by Rampino and Stothers, and was surprised by his strong reaction. Rampino, he told me, had claimed at the Snowbird Conference that the iridium layer was due to chemical precipitation in the oceans. Later he claimed that the iridium layer had been laid down by a volcanic eruption, without explicitly retracting his previous stand. Luie thought it almost funny that so soon after disagreeing with the impact theory of the Cretaceous extinctions, Rampino would write a paper that not only explained that catastrophe in terms of an impact, but also explained eight others the same way. "I bet he doesn't mention, let alone retract, his two previous theories," Luie said.

I decided I should not call Rampino and Stothers just yet. I had to try to figure out how they had solved the problems with galactic oscillations, problems that I had failed to solve. I had to do enough homework so I could appreciate any subtle arguments they might give me over the phone. I decided to give it a few days of work.

Walt invited me to drive with him and Luie down the San Francisco peninsula to Stanford to hear Thomas Ahrens give a lecture on the physics of the impacts that had caused the mass extinctions, but I declined. I had too many things to do at Berkeley. On the drive there and back, Luie and Walt talked about Nemesis, and who might find it. They talked about searches using visible light, and the possibility of using the data recently taken by IRAS (Infrared Astronomical Satellite). Walt decided that Nemesis was too big a plum to leave for someone else to pick. It should be found by the Berkeley group. Luie agreed. It was the obvious next step.

Just as I had decided I needed a real astronomer and had called Marc Davis, so Luie decided that *he* now needed a real astronomer, and his pick was David Cudaback. He was on the faculty of the Astronomy Department while simultaneously acting as an associate director of the Radio Astronomy Laboratory. Dave was in Hawaii, but after a telephone discussion with Luie, he flew back to Berkeley early. The next day he met with Luie and Walt, and they gave him copies of the two papers. The day after that, Walt and Dave began to hunt through the many bound volumes of star catalogues in the Astronomy Library, hoping to find a good candidate star for Nemesis. It was possible that Nemesis had been catalogued.

The Nemesis paper had not yet been published in *Nature*, yet many people at Berkeley were beginning to hear about the Nemesis theory and the crater paper through the scientific grapevine. I was asked to give a lecture, a Research Progress Meeting, at the Lawrence Berkeley Laboratory. These are usually talks describing research still in progress and not yet published. Judy Goldhaber in the public relations department asked if she could prepare a press release, but I said no. It is traditional in science that work should not be described in newspapers until the professional paper is published, in this case in *Nature*. This important tradition allows other scientists who hear about some new work to immediately read the scientist's detailed account of the work. Judy then suggested that she prepare a press release, but not release it. That way, if the newspapers somehow got hold of the story, we could at least give them a carefully thought-out version to work from. That sounded like a good idea, and I gave her permission to do it.

A few days later, I was called by a science reporter, David Perlman, of the *San Francisco Chronicle*. He asked, "Do you know this fellow John Hewitt who is trying to sell a news story about a companion star to the sun?" I had never heard of Hewitt. "He has copies of two papers you wrote," Perlman said, "and has written a story about it that he wants to sell to the *Chronicle*." Perlman had read both papers, and thought it would be a real "scoop" for the *Chronicle* to release the story. "I felt that I had to accept his offer," Perlman said. "If we don't buy it, somebody else will. But then I saw in the University *Bulletin* that you are giving a Research Progress Meeting tomorrow. Since you are talking publicly on this subject, I no longer feel any obligation to use Hewitt's version, which isn't very well written anyway. I'll be coming to your talk. Can I also have an interview?"

I found myself a bit confused by the paradoxical standard of newspaper ethics that I was hearing. Apparently Perlman could not write about my work without Hewitt, since Hewitt had brought the work to his attention, until Perlman discovered I was giving a "public" talk on the subject. I decided that authorship in newspapers is at least as complex an ethical question as authorship on physics papers.

I was in a bind. If I asked Perlman to delay publication until the paper was published in *Nature*, this Hewitt fellow would go and sell his story to another newspaper. An interview with Perlman would at least assure that he got all his facts right. I knew I would be criticized by other scientists for "publishing in the newspapers." Most people feel honored when they get their names in the paper, but I had had enough of this "honor" in the past so that it no longer brought any thrill.

The best thing, I decided, was to allow Perlman to come to my talk, especially since it wasn't clear that I could stop him, and then give him the requested interview. I could try to atone for the sin in the eyes of my colleagues by making sure that Perlman referred to the work of Whitmire and Jackson, the two scientists who had independently come up with the companion-star theory. I told Perlman about their work, and also gave him their phone numbers. I also suggested he call Rampino and Stothers, to hear what they had to say about galactic oscillations, although I had not

yet understood how they had made that model work. He agreed to talk to
them and include them in his article. My presentation was scheduled for
the next day, Thursday. I asked him if he could delay his publication a
little, so that I could call everybody first and warn them of what was going
to happen. He said he would hold his article until Monday provided I
promise to alert him if any other reporters called me. I told him that
Goldhaber had prepared a press release, but we wouldn't distribute it until
Monday, the same day his article would appear.

Then I called up everyone who was involved. Piet Hut was out of town,
but I reached Marc Davis. He wasn't too surprised that the story had
leaked. When I told Walt, however, he was so furious that he tracked
down John Hewitt and called him. He learned that Hewitt had once
worked in the Astronomy Department, and after leaving had continued to
visit the library there. At the library he had found copies of our papers,
which we later learned had been left behind by Cudaback so that interested scientists could read them. Hewitt had decided that this was his
opportunity to break into popular science writing. He hadn't done anything wrong. It was inevitable that if we distributed copies of our paper
before publication, the newspapers would get hold of them.

I called Whitmire and Jackson, explained to them what had happened,
and told them to expect a call from Perlman. They seemed not to be
particularly bothered. I then called Rampino and Stothers; Stothers
wasn't in, and I spoke only to Rampino. I told him about our work, and
how I had heard about his from Dave Raup. Then I told him about the
Chronicle story that was being written. He was outraged, even more than
Walt, but unlike Walt, he seemed to be blaming *me*. I apologized and
tried to explain that it wasn't really my fault and that I wanted to make
sure he received proper credit in any articles that were written. He said
that he would call *Nature*. I told him I would send him copies of the two
Berkeley articles, and I asked if he would send me a copy of his paper. He
refused. He said that he thought it was inappropriate to distribute copies
of papers that have not yet been accepted for publication.

I gave my talk on schedule the next day, and Perlman sat in the back
taking notes. Afterward he and I talked for about a half hour. I was
pleased to find that he had understood most of the technical details of the
talk, and I expected that his article would be accurate.

The following day I received a call from an editor at *Nature*. Rampino
had spoken to him, and he wanted to make sure that I knew their policy:
Any paper that is publicized before publication will not be printed. I

agreed with the need for such a policy, but emphasized that this publicity was not done on purpose and I hoped that they would not penalize us. Secretly, I felt confident that they wouldn't, but I didn't say so on the phone. I was quite deferential. The editor admitted that the leak was not my fault, and he thought that they probably would publish our papers anyway. He said that they would put our two papers in the same issue, along with those of Whitmire and Jackson and Rampino and Stothers. I asked him whether there were any other papers that had been submitted to explain the periodic extinctions; the editor said no.

That weekend an article in the *New York Times* caught my eye. It was titled "Earth Believed Facing Comet Bombardments." Had Hewitt sold his story to the *New York Times*? Or had somebody else independently come up with the Nemesis theory? No. The story was about the work of Rampino and Stothers at Goddard. I read quickly through the article, hoping it would help me understand how they had solved the problem that I had failed to solve: to make the extinctions synchronize with some aspect of the galactic oscillations. The article seemed to say that the extinctions were caused by comet showers, triggered when the sun crossed the galactic plane. But the phase was wrong, I said to myself. The reporter, Walter Sullivan, had perceptively asked Rampino the critical question: When would the sun next cross the galactic plane? Rampino's reply was accurate: "It's there right now." Rampino went on to say that the molecular clouds that trigger comet showers are scattered randomly above and below the plane, so the close encounters are spread out over a long period. Everything he said was true, but it didn't answer my objection. I was still mystified by their theory.

There was an article on the same page with the headline "Labor Leaders Weigh Impact on Campaign." It took me a moment before I realized that it wasn't the impact of a comet they were talking about.

It turned out that the reason I had been unable to reach Piet Hut at Princeton was that he was in Berkeley to give a lecture for the Astronomy Department. On Monday morning he was having a cup of coffee when a friend brought in the morning *San Francisco Chronicle*. He was surprised by a story with the headline "Death Star Kills Life on Earth," by David Perlman.

The story described our work in detail, and it also cited the work of Rampino and Stothers and of Whitmire and Jackson, although the mentions were briefer than they should have been. The article was picked up by the wire services, so Perlman had his scoop despite the preceding article in the *Times*. Walter Sullivan called me that afternoon to find out more about our theory. After talking with him for a while, I asked him where he had gotten the Rampino and Stothers story. "They called me on the phone last Friday. I met with them, and they gave me a copy of their paper," he replied. I asked Sullivan what they had told him about the work done by me and my colleagues, or by Whitmire and Jackson. "Nothing," he replied. "I first heard about it from the wire services." Sullivan wrote a second article, describing the companion-star theories, in "The Week in Review" section the following Sunday.

Cudaback's search in the existing star catalogues was beginning to look fruitless, and Luie had begun thinking about other ways to find Nemesis. What kind of experiment, what kind of observation program should be set up? I don't know how many ideas he had before he found one he thought would work. He stopped by my office to tell me about it.

"Nearly every star in the sky is moving toward or away from the sun at a velocity of a few miles per second or more. Nemesis is right now about as far away as it can get, and it is just turning around, beginning its fall back toward the sun and the Earth. That means it is one of the very few stars in the sky that is not moving toward or away from us." Luie continued, "We can find Nemesis by looking for a star with nearly no red shift."

He was referring to the fact that objects moving toward or away from us show a shift in their color by a tiny but measurable amount. If we measure the color of certain features in the star (such as hydrogen emission lines) with high accuracy, we can determine the star's radial velocity. The change in color is often referred to as red shift, because most galaxies in the Universe are moving away from us, and that makes their colors redder, although certain galaxies and stars show a blue shift. Nemesis would have virtually no shift, red or blue.

Unfortunately, red shift is not particularly easy to measure, especially if you have to do it with a half-million stars. However, Luie had invented a way to make the survey possible. He described to me a new kind of optical filter, which had a very narrow color acceptance tuned to an emission wavelength of hydrogen. If one of these was inserted in front of the photographic plate, the only light coming through would be from stars with zero red shift. It was a Nemesis filter, cutting out the light of nearly every other star.

It was a clever idea, but I immediately saw problems. The light from

Nemesis was already expected to be quite dim. I had calculated that it was about a factor of 100 too dim to be seen with the naked eye. Luie's filter would reduce the brightness of the star even further. "That's no problem," Luie insisted. The filter would make Nemesis dimmer, but it would make nearly every other star black. "Except for the continuum," I argued, pointing out that many stars have a broad spectrum of light, and do emit *some* light at the frequencies expected for the Nemesis spectral lines. Luie thought this wouldn't be too much of a problem, and we spent some time talking about it without reaching a definite conclusion.

Another problem, I decided later, was that we didn't know the position of Nemesis as accurately as Luie thought we did. It might not be at the end of its orbit, just turning around; it might be only halfway out. If we chose the wrong frequency for Luie's filter, it would cut out Nemesis as well.

The real effect of Luie's invention was to wake me up. I had already begun to be lazy again. I was so proud of the Nemesis theory, and the crater work with Walt, that I had almost stopped thinking about new steps to take. Luie and Walt were absolutely right about one thing: The next step was to find the star. Why hadn't I tried to invent a method? Why had I left the problem to Walt and Luie and Cudaback?

The key to finding Nemesis, I decided, was in its closeness to the Earth. The stars nearest to the sun had all had their distances measured by using the fact that their apparent position in the sky changes slightly during the year. As the Earth moves around the sun, we look at the nearby stars from a slightly different angle. Photographs taken three months apart would show that Nemesis had moved slightly with respect to the more distant stars in the photos. The slight shift in the apparent position of the star is called its parallax. Stars with a parallax of one second of an arc are about 3¼ light-years away. This distance is also referred to as one parsec, the distance at which the parallax is one second. I always thought that the parsec was a bad unit to use in astronomy, and it annoyed me that it was so popular. It seemed to me very artificial to measure distances in relation to the diameter of the Earth's orbit. It was ironic that the distance we predicted for Nemesis should turn out to be so close to one parsec.

Since Nemesis must be closer than any other star, it should show a larger parallax than any other star. Sets of photographs taken three months apart would show Nemesis shifted by almost one and a half seconds of arc. After another three months, the shift would increase to nearly three seconds of arc, before it began to decrease again as the Earth turned in its orbit. Three seconds of arc is still a small shift, not something that could be easily spotted by looking at photographic plates. I thought about trying a method that had been used to look for new planets and asteroids, in which the two photographs taken at different times are flashed alternately on a screen. If one object has moved, it will seem to jump back and forth slightly, while everything else remains fixed. The eye is very good at noticing such a small movement, for reasons that probably have to do with primitive man the hunter, who, in looking at a scene, tried to spot the camouflaged prey moving slightly against the background of the forest. Or maybe the reasons have to do with primitive man the prey, looking toward the woods and hoping to spot the camouflaged tiger before the carnivore spots him. Whatever the reason, our eyes seem to have built-in computers excellent for spotting motion.

The only practical use I had previously made of that knowledge had been in attracting the attention of friends in a crowded room. It is much more effective to wave your hand slightly than to stick it up in the air. The only known creatures on Earth who seem to have lost the ability to spot a waving hand belong to a subspecies known as waiters. Some scientists believe their inability to spot such motion is learned rather than inherited.

The problem would not be in spotting the star, but in taking all the photographs. Each photo (or plate, as they are called in astronomy, because they are mounted on a glass plate, for rigidity) must be made at high magnification and with great care to make sure that it will not become distorted as it is developed, and thus lead to a false alarm. Thousands of plates would be necessary to cover the whole sky, and special instruments would be needed to compare each section of the plate with its partner taken months later. It sounded expensive and tedious. I wondered if there was a better way.

Walt called me to let me know that Gene Shoemaker was visiting Berkeley to talk with him and to give a lecture on Earth impacts. He was the scientist who had correctly told Raup that my original narrow orbit was "unstable." Shoemaker was one of Walt's scientific heros, and he was anxious for me to meet him. Walt called him a "geologist's geologist." He said that Shoemaker not only was an excellent field geologist, but also had played a major role in extending the study of geology to encompass the other planets. He was the first scientist to gather convincing evidence that Meteor Crater was from an impact. "He stole the planets from the astronomers and gave them to the geologists," Walt recalled someone as saying. He probably would have been the first scientist astronaut to explore the moon, but had been eliminated because of a minor medical problem. He contributed to the moon program by training all the astronauts in geology.

Gene had been looking at our Nemesis theory and the crater analysis, and had told Walt that we were wrong. Walt said that Gene would undoubtedly repeat his remarks in his lecture, assuming that he thought our theory worthy enough to take the time to criticize. I was very concerned and curious, but unfortunately I couldn't attend his lecture; it was being given at the same time as a course in quantum mechanics that I was teaching. Walt suggested we get together in the evening at Luie's home to see if we could defend ourselves. Marc Davis was away on his sabbatical, but Piet Hut was in town, and I invited him along.

That evening I met Gene Shoemaker for the first time, and I was pleasantly surprised. Despite all the good things Walt had told me about him, I think I must have expected a stuffy curmudgeon who would pontificate about his expertise on the solar system. I have a natural awe and fear of true experts, so I always imagine them as ogres. Instead, I discovered that Gene was one of the most charming and friendly men I have ever met. He greeted Piet and me like long-lost friends, and told us a few jokes. I began to regret having missed his seminar.

After discussing everything from the weather to the pleasures of geology, we began to get down to business. Gene's first objection to our theory was that there already were enough asteroids and comets in space to account for all the known craters on the Earth and the moon. Our additional comet showers would make too many extra ones. Walt argued with him on this. The number of comets and asteroids presently out in space was uncertain, he claimed. There might be only half as many as we think, since most of the asteroids Gene was speaking about had never been observed directly. They were too small. Their presence had been estimated from the craters they had made; maybe those craters had been formed from comet showers rather than from unseen asteroids. Gene countered that, if anything, we have probably *underestimated* the number of asteroids out there.

I sat back, not openly taking sides, as these two geologists discussed this difficult question. I felt that Walt must be right. I had seen the estimates for the number of asteroids change by a factor of two in just the last several years. To estimate the number of asteroids of a given size, one must guess what their surface reflectivity is, since the total reflected light is the only estimate we have of their size. They are too small to appear as disks in ground-based telescopes; rather, they look "starlike,"

like points of light. Anything that depends on guesses, and that changes from year to year, cannot be too well known.

Gene didn't make any progress in convincing us that the present value for the number of asteroids out there was accurately known, so he brought up his other objection to the Nemesis theory. He explained to us that in the orbit that we had hypothesized, Nemesis would be stable for only a few million years, and then it would get ripped from the sun by a passing star. It was just the kind of argument I most feared. I had become an expert on the stability of the orbit shape, but not on the orbit lifetime. I had relied heavily on Piet's theoretical expertise, perhaps too heavily. Thank goodness Piet was there, and we might be able to find out who was right.

Piet immediately disagreed with Gene, and he began giving a very simple model to demonstrate why he felt the lifetime of the star would be closer to a billion years, rather than to Gene's claim of a million. Passing stars would each give a nudge to Nemesis. The biggest effect would come, Piet argued, from close encounters with big stars. We could estimate the impulse by the time it takes the star to pass. Since stars are typically moving about 30 kilometers per second, and since they influence Nemesis significantly only while they are within about half a parsec, Piet had calculated that the time of passage was about 30,000 years. About a million years would pass before another star went by, and during this time the sole influence on Nemesis would be the sun. Thus the effect of the sun lasts about thirty times longer than the effect of the passing star. Moreover, the effect of the sun is always approximately in the same direction, whereas the effects of passing stars tend to be random and tend to cancel each other. So for the effect of numerous passing stars to be equal to that of the sun, so they could gradually rip Nemesis away, they would need more than thirty passages. Using the mathematics appropriate to such "random walks," Piet had estimated that it would take $30 \times 30 = 900$ passages. At 1 million years each, that would mean the orbit would be stable for nearly a billion years, just what we had claimed in our paper. I was relieved. Piet's argument had been simple enough for even me to follow.

Of course, Piet went on, a billion years is not long enough for Nemesis to have lasted in that orbit since the beginning of the solar system, which, we think, was about 4.5 billion years ago. Nemesis must once have been closer, as we'd said in our paper, and had now worked its way out to the present orbit, where it has about a billion years left to live. You can show, Piet continued, that the gravitational binding energy of its orbit decreases linearly to zero as time goes on. That means that 4 billion years ago it must have been about four times closer to the sun than it is now, assuming that it was following the average behavior. Since Nemesis is a unique object, however, its history could have been different.

I was anxious to hear Gene's response. To my disappointment, he simply said that he thought Piet was wrong and that he would run some computer simulations to demonstrate it. I was unsatisfied with this response, so I began to press the case. I asked Gene to point out which step in Piet's logic was wrong. Gene simply refused to do so, and once again said that the computer would show who was right. I suspected that maybe he had not followed all of Piet's argument; unlike me, he had not heard it before. Or maybe he had followed it and was just bothered by the fact that it gave an answer different from what he had expected. Luie and Walt were silent. Then Piet agreed with Gene that a computer simulation would be the ultimate answer. I wasn't sure whether he was just letting Gene off the hook or if he really felt that the simulation had some chance of giving a different answer. I found it hard to believe that Piet could be wrong by more than a factor of two. Certainly Gene's claim of a million-year lifetime, a thousand times shorter than the one Piet had calculated, could not possibly be right. Gene said that he had a program already running back at the U.S. Geological Survey in Flagstaff that would do the job, and he would telephone his lab in the morning and ask his colleagues to run it. We could know the answer very quickly. Piet decided that he could easily modify his existing computer programs back in Princeton and find the answer himself. I offered to pay his round-trip airfare to Princeton out of my research grant, and he left for a week of day-and-night computer work.

The next night, Walt took Gene out to dinner at one of the excellent Chinese restaurants in Berkeley. Several of Walt's graduate students and colleagues went along. Afterward Walt told me that a lively discussion had ensued over whether "Rich's star" could really be up there or not. No progress was made until the end of the dinner, when the fortune cookies were served. To the amazement of everyone present, Walt's fortune read, "The star of riches is shining upon you."

14. Conference

THE NEWSPAPERS loved the Nemesis story. It gave them an opportunity to revive the stories of the destruction of the dinosaurs, of the Alvarez impact discovery, and of the nuclear winter, as well as the fascinating claims of Raup and Sepkoski that such catastrophes occur on a regular schedule. But new features had been brought in: comets, craters, and the death star dubbed Nemesis. The story was carried on both the CBS and NBC national evening news programs. It was spread around the world by the Associated Press and United Press International. Feature articles were written in the *Los Angeles Times*, the *Washington Post*, *Discover* magazine, *Science*, *Science 84*, *Science Digest*, *Sky and Telescope*, *New Scientist*, and others. I was inundated with phone calls. At one point I couldn't take a long-distance call from Australia because I was live-on-the-air on a Salt Lake City talk show. *Scientific American* wrote a short column entitled "A Star Named George," from one of the alternative names that we had suggested in the preprint of our paper. Stephen Jay Gould liked the theory. It seemed to fit in very nicely with the "punctuated evolution" idea he had created with Niles Eldredge. But he suggested in *Natural History* that the star should be named "Siva" instead of Nemesis, not knowing, of course, that we had already rejected a variant of this name. The rock group Shriekback included a song called "Nemesis" to start off the second side of their album *oil and gold*.

Walter Alvarez hated the publicity. He felt that it interfered with his research and didn't do him any good. I didn't particularly mind the attention, although I was annoyed at the inaccuracy of most of the stories. The public loves science, and yet scientists don't usually do a very good job of letting them know what is going on. Anything really exciting in science should be communicated to the public, I thought, and, besides, it was fun doing so. I enjoyed trying to show how the complex theories really weren't that difficult to understand. Even Congress wanted to hear about Nemesis. Alvin Trivelpiece, the director of research for the Department of Energy, telephoned me at home one evening. He was about to testify before Congress on the research supported by his department, and he wanted to make sure he had all the facts straight.

It was only months later that I began to realize the harm that was done by the publicity. Every science reporter who wrote about Nemesis had local scientists whom he depended on for guidance, comments, and perhaps some color and controversy. Thus many scientists were asked to comment on the new theory, but almost none of them had yet seen our papers. As a result, they often said things that were either irrelevant or wrong. "I would think an orbit like that would be unstable," a prominent scientist at Harvard was quoted as saying. He never would have said that in a published paper without putting a great deal of thought into it. In a conversation on the telephone people are more casual.

Even worse were those scientists who jumped knowingly into the fray and wrote papers opposing our model before they had a chance to read ours. In the next several months I saw several published papers that had basic facts from our paper all wrong, or confused details of the Nemesis paper with details from the paper by Whitmire and Jackson. (They had a much smaller companion star, visible only in the infrared and having a more eccentric orbit.) When colleagues asked me whether we had any response to the "latest criticism," I often responded by saying, "Yes, and you can read it in our original paper."

The idea of a conference began in a conversation between Dave Raup and Luis Alvarez, who was now enthusiastic about the Nemesis theory. Walt, Frank, Helen, and I all agreed that it would be exciting to hold a small conference at Berkeley, with all the major contributors attending. Frank and Helen volunteered to organize it. Dan Whitmire and Al Jackson readily accepted the invitation, as did Gene Shoemaker. Dave Raup and Jack Sepkoski flew in from Chicago; Piet Hut returned from his week of orbit-stability simulations on the Princeton computer. Leading paleontologists and geologists accepted invitations, including Erle Kauffman, Gerta Keller, and Al Fischer, who had finally been vindicated in the ideas he had presented on periodicity in geology some years earlier. Rampino and Stothers declined our invitation, despite our offer to pay their travel expenses. Luie felt that this conference would be truly historic, perhaps looked upon someday as a turning point in our understanding of the solar system, the Earth, and even evolution. I found it hard to take his grandiose vision seriously, but I was also aware that Luie had much more historical vision than I had, having lived through much more history.

The conference was held on the weekend of March 3 and 4, 1984, at the Lawrence Berkeley Laboratory. I finally met the people behind all the papers. Most of them turned out to be younger than I had expected; maybe I turned out to be younger than they expected. Raup was clearly the established, senior member of the Raup and Sepkoski twosome. Sepkoski was tall, athletic, and bursting with energy. He exuded a boyish enthusiasm and excitement. Whitmire and Jackson seemed pleased to have Nobel laureate Luis Alvarez pay them so much attention and praise their work. Shoemaker, as the senior geologist of the crowd, the one who knew about craters and asteroids before anybody else in the room had shown any interest, was often called upon for his experience and expertise. He used his charm adeptly in preventing disagreements from turning into arguments.

We had peppered the conference schedule with short breaks, since we knew that a lot of what is really accomplished at meetings is done informally during these times. I spent a considerable amount of time with Whitmire and Jackson. Whitmire wanted to try out a new idea on me, the possibility that a tenth planet could cause the periodic extinctions. He called this hypothesized planet "Planet X." The X had a clever double meaning. It stood for "unknown," as well as for the Roman numeral 10. If the orbit was in the right place, and precessed just so . . . I found it difficult to understand Whitmire's speculations, and I decided he was off on the wrong track. Months later I found out how wrong I was, and how clever this new idea of Whitmire's really was. It is amazing how quickly one can become pigheaded.

Whitmire said that they hadn't yet put together a final version of their companion-star paper; they were waiting to see what they might learn at this conference. I told him that a delay on their part might delay publication of their paper, and perhaps give ours a priority that it didn't deserve. I offered the services of my secretary and a word processor. That evening

Whitmire and Jackson worked late, preparing the final changes in their paper to send to *Nature*.

I learned just as much listening to my Berkeley colleagues as I did from hearing the formal talks of paleontologists and geologists. It takes hearing an organized talk to understand someone's full approach to a subject. Everyone at the conference seemed to have a different perspective on the farrago of problems we were addressing.

The talk I found most interesting was the one given by paleontologist Erle Kauffman, who introduced himself by explaining that he had once been "a traditional gradualist." He obviously hadn't completely given up the old religion, because he spent much of his talk showing data that demonstrated that mass extinctions were drawn out over extended periods of time, in contradiction to the prediction of the impact theory. Luie had always told me that the paleontologists must simply be wrong in this claim, perhaps because they didn't really understand statistics, but as I listened to Kauffman, I found it difficult to agree with Luie. Kauffman showed that there had been a great catastrophe in the coral reefs a half-million years before the dinosaurs disappeared. A half-million years is not a short time, even for geologists. Ten meters, more than 30 feet, of sedimentary rock can be laid down during that period. He showed slides that indicated that extinctions of different families had really occurred at different times. I couldn't find any valid reason to dismiss his extensive data.

Then I remembered that Luie's belief in the suddenness of the extinctions had predated the Nemesis theory. We no longer expected a single impact, but a shower of impacts spread out over a million years or more. Some species might disappear with the first impact, others with the second, and the most robust (such as the dinosaurs) with the third. A paleontologist might well interpret this as a drawn-out period of crisis, as an extended extinction.

It all seemed so clear. The paleontologists had never been willing to back down from their theory of gradual extinctions, not because they were stubborn, but because they were right! But Alvarez had been right, too: Catastrophes had been caused by impacts. They had both been right. No one had imagined that there was a possibility of reconciling these two, apparently contradictory, stands. The concept of the comet shower had provided the solution. This was *harder* than a jigsaw puzzle. At least with a puzzle you know what pieces you are looking for. In this business there seemed to be no missing pieces until we found them.

Science doesn't work in the logical way that it should. The clues are too fuzzy, too easily dismissed. It is hard to find the right theory when the facts are still in dispute. Four years earlier, we should have said: "The paleontologists have shown the extinctions are extended. The Alvarez collaboration has shown the extinctions were caused by impacts. These are two established facts. How can we reconcile them?" We might have concluded that the impacts occurred in bunches, and thus been led to the idea of comet storms, even before the periodic extinctions had been found. But instead of looking for a new idea that would solve the paradox, reconcile the evidence, years had been spent trying to see which of the two bits of evidence was wrong.

After Kauffman's talk, I got up and explained that with the new comet-shower model there was no longer an inconsistency between the view of extended extinctions and the idea of impacts. There was almost no reaction to my comments. I realized afterward that many of the paleontologists felt impacts were only part of the cause of the mass extinctions, so they saw no need for a reconciliation. Neither did Luie. They weren't ready to see that everything now fit together. Paradigms in science are not easily changed.

Piet Hut gave a beautiful talk on the Nemesis orbit, explaining why it was sufficiently stable to give regular comet showers over the last 250 million years. He had done extensive computer simulations, and they had all verified his simple, but accurate, physical reasoning. Listening to his clear presentation, I felt proud to have such a person as a coauthor.

Shoemaker was next. He almost knocked me out of my seat with his opening remarks. "I would like to apologize to the Berkeley group," he said. "I was wrong when I said that their orbit would be broken up by passing stars after a few million years." His computer simulations showed that Piet's theory for stability had been correct; the correct lifetime for the present orbit of Nemesis was closer to a billion years. He had previously given several talks dismissing our results, but he now admitted that he had been mistaken, not us.

Luie had taught me that public retractions were a necessary step when you find that you have been mistaken, but I had never imagined that an apology was necessary! I wondered if I would have had the courage to do the same thing.

His apology over with, Gene was careful to point out that he still had serious problems with the Nemesis theory. For one, there were too few craters on the Earth to accommodate all the comet showers. It was the same point he had made to us at Luie's house the evening I first met him. Second, the extinctions were not really periodic, as Raup and Sepkoski had claimed, but just random in time with an appearance of periodicity. And, finally, the a priori probability of finding a companion star was so small that it was hard to take the Nemesis theory seriously. He pointed out that there were no known binary stars with separation comparable to that of the Nemesis-sun system. Actually, he never used the name Nemesis. He referred to it as the "putative death star," thereby emphasizing his disbelief.

Luie took on his probability claim. The probability of the star's being there must take into account the fact that we are here, Luie argued. Perhaps periodic extinctions are necessary to give complex life forms, such as humans, a chance to unseat the dominant and primitive species that occupy all the ecological niches. I didn't completely agree with Luie's reasoning, since it seemed to me that the same reasoning could be used to bolster any highly unlikely theory, but it was certainly an intriguing point.

Walt next addressed Shoemaker's comments about the compatibility of the model with the number of craters. He had done some homework since our last meeting with Shoemaker. He gave a detailed breakdown of the uncertainties in the estimates of the number of asteroids actually out there in space, using Shoemaker's own published papers. He showed that the uncertainty was great, and easily accommodated the factor of two or three that we needed. Only half to two thirds of the craters on the Earth are formed during showers in our model; the other half to one third could come from the asteroids that Shoemaker had measured.

After Gene's attack on the periodicity, Raup and Sepkoski defended their work, but they did it too weakly, I felt. So I got up and showed the plot I had prepared with my Fourier analysis of their data. None of Gene's arguments applied to my method of analysis, I said, and yet the 26- and 30-million-year periods showed up strongly in my plot. Gene didn't really answer, but I had the feeling that I hadn't fully convinced him. (I was surprised to learn, much later, that his real response was to go back home to Arizona and redo *my* method of analysis, to see what potential flaws, if any, lay within it. He was not about to criticize me just because I disagreed with him. He wanted to take the time to understand my methods in detail. It was an approach to science that I had rarely had the patience to take, but perhaps I had more to learn.)

Frank Asaro gave a talk on the iridium measurements. He had been slowly and carefully accumulating measurements, and now had an impressive collection of data. The plot that I found most interesting was the one that showed the possibility of multiple iridium peaks near the end of the Eocene, just what our theory would predict. Frank was very cautious, however. The multiple peaks could be spurious, caused by mixing of the sediment. The final word wasn't in yet.

Frank also showed his measurements of the Cenomanian-Turonian mass-extinction boundary, the one that preceded the Cretaceous by 30 million years. An Italian group led by Forese Wezel, a geologist at the University of Urbino, had reported abundant iridium at this boundary, but Frank had found none. He showed that the Italian samples had probably been contaminated. Too bad. It always hurts a little to see data that support your theory go away. (Later Frank would also disagree with reported measurements made by a Chinese group showing an iridium level at the Permian-Triassic boundary. We ultimately found iridium near the Cenomanian-Turonian boundary, but not where Wezel had reported it.) Frank was one of the few people in the world whose judgment I could trust without resorting to double-blind techniques. (I don't include myself in that list.) He didn't care how much some result supported his own work; it had to be carefully checked, and if it was wrong, it was wrong.

Luie gave a presentation in which he showed a new type of iridium detector he had invented. It would eliminate the long, tedious chemical preparation steps that had severely limited the number of samples that could be measured each year. The detector would make use of the fact that neutron-activated iridium, Ir-192, often emits two gamma rays in its decay. About half of the decays would have both a 468 keV gamma followed (in less than one ten-billionth of a second) by a 316 keV gamma. (KeV, kiloelectron volt, is a unit of energy; a typical medical x-ray has an energy of between 50 and 100 keV.) The new detector would cost a few hundred thousand dollars to build, but it could be used to measure 20,000 samples per year. Sedimentary rock dating from the present to 250 million years ago could be sampled every centimeter and measured with this new device. It would be the ultimate machine to see how many iridium layers there actually were in the geologic record.

Frank had argued that you lose sensitivity if you require detection of both emissions, but Luie had countered that the lost sensitivity would be more than compensated for by decreased background emissions. They had bet a dollar on this issue: Luis Alvarez, the great inventor, challenged by Frank Asaro, the great expert in neutron activation. I couldn't guess who would win.

(A year later, Luie won the bet, when Frank completed the detector, improved it, and made it work just as Luie had predicted. Frank was delighted to pay. In a sense, he had been the smarter of the two. He had known that he couldn't lose; he would get either a better detector or a dollar.)

Everyone at the conference was disappointed that Rampino and Stothers had not come. During a coffee break I mentioned to Richard Kerr, a scientist turned reporter working for *Science*, that they had refused to send me their paper. "That's curious," Kerr said. "They sent *me* a copy. I have it here." He gave me the paper, and I quickly read it.

I searched through it for the idea that I had missed, the idea that made the galactic-oscillation picture work. I didn't find it. I searched again. They didn't say anything that I didn't know about. But they had done a detailed calculation of the galactic crossing times. Maybe that was the key. I looked at their table of galactic crossings and compared it to their list of mass-extinction times. They said the agreement was "excellent," so good that the probability of its happening from chance was less than 1 in 1,000. They had found, as had I, that the sun is crossing the galactic plane right now, yet the last extinction was 11 million years ago. Well, one disagreement doesn't kill a theory. The sun crossed the plane 65 million years ago, right during the Cretaceous catastrophe. But one agreement doesn't make a theory either. I looked at the other dates. There was very little agreement. There were many misses, by as much as 13 million years. You can't get a miss bigger than that, I thought, because you would simply assign the extinction to the *next* crossing. There seemed to be no correlation whatsoever between the two lists of dates. Had they made some horrendous mistake?

I prepared a little plot, showing that there was absolutely no correlation between the galactic crossings and the mass extinctions, and asked for five minutes to show it at the conference. Their theory was simply wrong, I said, much relieved. They had made a mistake. I didn't know how they had calculated the odds of 1 in 1,000 of their correlation's being accidental, but they were clearly wrong. I wished that they were there to defend themselves against this criticism. I would have loved to see them try.

A few weeks later, I received a copy of a letter from a statistician at Chicago, S. M. Stigler, who had figured out exactly what mistake they had made. Stigler's letter was eventually published in *Nature*. In comparing the two sets of numbers, Rampino and Stothers had calculated a "correlation coefficient." They then had compared this correlation coefficient to a

theoretical value expected for random numbers. But in doing the compar-
ison, they had forgotten the fact that their numbers (the dates) had been
arranged *in order*, from oldest to youngest. Their theory assumed that
the numbers were unordered. Any two ordered lists will show a "correlation" using this technique. The correlation will be even stronger if the
events in both lists are roughly equally spaced. The correlation coefficient
ignores the relative phase of the two data sets, i.e., whether the events in
the two lists actually occur at the same times.

It had been a simple, but devastating mistake. Rampino and Stothers
eventually wrote a letter to *Nature* attempting to refute Stigler's criticism,
but I decided that Stigler was clearly right. Several months later, I finally
met Rampino, at a conference in Arizona, and I showed him my analysis
that showed the correlation was negligible. He didn't back down, even
though he admitted he had no answer to my criticism. He said that it was
Stothers who had done the statistical analysis, and he had just accepted it
all, but he believed in Stothers, who was an expert. I gave him a copy of
my plot to show Stothers. I never heard from him again or from Stothers
on this issue.

The next presentation at the Nemesis conference was given by Dave
Cudaback, who described his search for Nemesis in the existing star
catalogues. As I sat there listening, I started to think, once again, about
finding the star. The trick, I still thought, was to take advantage of the
large parallax. We would have to take thousands of photos. Then I
thought about another project of mine, the automated supernova search.
Physicist Carlton Pennypacker and graduate student Jordin Kare had
taken over most of the responsibility for that work, since I was distracted
by this dinosaur business. Over the last several years they had helped put
together a system for looking at thousands of galaxies every night. It
wasn't quite working yet, but it was getting close. Instead of film, we used
a television camera (made by us), which fed the image information
directly into a computer. The computer also pointed the telescope, allow-
ing us to look at a new image several times each minute. How long would
it take to survey the entire sky for Nemesis?

I quickly calculated that the sky has a total of about 40,000 square
degrees. Suppose we took an image of each? We would have to measure
parallax to one third of an arc second. That would mean 1 part in 10,000
across our imaging plane. It would be hard, but I thought we could do it.
If we could make 1,000 images a night, the sky would be covered in a
little over a month. Then three to six months later we could take another

1,000 images. The computer would look for a large parallax, and spot Nemesis. We could survey only the Northern Hemisphere, but that gave us a 50% chance of success. We could do the Southern Hemisphere later.

The idea looked good, very good. Why hadn't I thought of it before? We had a computer-driven telescope, virtually a robot. Was the idea right? I checked the numbers, and they looked okay. I began to get excited. Luie, sitting next to me, was chairman of the session. As Cudaback was finishing, I whispered to Luie, "I have a new idea I would like to tell the group about. May I have five minutes?" Luie gave me the time. I was taking a chance. An idea that is only five minutes old probably will live for only five minutes, I remembered. But how could this simple idea be wrong?

I went to the podium and showed my calculations to the gathered group. Nobody raised any objections, or pointed out any mistakes. Shoemaker found the scheme very interesting, and said he had other applications for our automated astronomy system. Afterward Luie told me that my search scheme was the most exciting idea he had ever heard of. I was pleased with the compliment, but accustomed to Luie's exaggerations. When he gets enthusiastic, he really gets excited. Later that day, I told him, "Luie, Nemesis is a good idea, but it will never live up to the discovery you and Walt made. You opened a new field, *this* field. Everything else is just building on the foundation you laid."

Luie disagreed. "No, Rich. Our discovery was important. But it is nothing like yours. If you find that star, it will be the most important discovery in astronomy, in science, in many, many years. You know the excitement that the discovery of Pluto caused. Nemesis is far more important to science than Pluto."

It suddenly occurred to me that maybe, just maybe, Luie was right. But I was too deeply immersed in the whole affair to be able to have any perspective, and too busy to spend time trying to develop some.

The only thing left to do was to find Nemesis, or so I thought.

15. Geomagnetism

O NE DAY in 1984 Dave Raup sent me a preprint that left me horrified. He had looked at some old data on the Earth's magnetic field and analyzed it in the same way he and Sepkoski had analyzed mass extinctions. He found that the rate of magnetic reversals was periodic, with a 30-million-year period. Dave Raup is a *nut*, I thought.

There was no way to link geomagnetic reversals to impacts. The magnetic field of the Earth comes from deep in the core, whereas impacts can affect only the outer surface, the crust and atmosphere. Linking the two was patently absurd. What really left me frightened was the thought that I had misevaluated Raup. Raup and Sepkoski had studied vast amounts of extinction data. They had to apply refined judgment to these data to decide which were good and which were inaccurate. Most of the data had been discarded. The remaining data had shown periodicity, the periodicity that had excited me two years earlier and that had eventually led to the Nemesis hypothesis. How could I be sure that they hadn't selected the data in such a way that the periodicity was artificially generated? They hadn't done the selection in a double-blind fashion, so there was no way to be sure. I had to trust them. Raup had been elected to the National Academy of Sciences, one of the highest honors a scientist can receive. Luie and Walt had spoken highly of him. His papers had seemed careful. He hadn't expected to see periodicity when he made the selection. At least that was what he said.

But now this ridiculous paper on magnetic-field reversals! He seemed to have the ability to find periodicities even where they didn't exist. My whole research program was now based on his discovery of periodic extinctions, a "discovery" that I could no longer trust. It was disconcerting, to say the least.

Now, be careful, Rich, I thought. My dismissal of his new work reminded me of Luie's original dismissal of Raup and Sepkoski's periodic-extinction paper. My major contribution back then had been to take their work seriously. Now that I was deeply into the extinction business, was I falling into the trap that Luie had once fallen into? Was my level of skepticism too high? Since things had come to make so much good sense to me, was I being too quick to ignore a new discovery? Was I now entrenched in my own paradigm?

I read Raup's paper carefully, looking for the fatal flaw. His analysis was very similar to that in his previous paper, although the statistics weren't quite as good. His mathematical confidence in the reality of the effect was only 97.5%, not 99.9%, as it had been for the mass extinctions. I decided that 97.5% wasn't good enough. That was sufficient grounds for ignoring his new paper. I felt somewhat relieved, and tried to forget Raup's crazy new result.

My conscience wouldn't let me forget it. How I would hate myself if I missed something important because in my old age (I was forty) I was becoming too conservative. I remembered the lessons I had articulated from my experience with Nemesis. It is easy to be a skeptic, to find excuses for dismissing new discoveries. If you are skeptical, you will usually turn out to be right. Most new discoveries are wrong. But . . . if you are *too* skeptical, you get left behind. You miss important implications. You do not discover new truths.

Why do new bits of evidence have to be so fuzzy, so marginal, so easily dismissed? Why couldn't things be clear and simple? Why couldn't Raup's evidence have been 99.9% instead of just 97.5%? ("Because then someone else would have found it earlier," I answered myself.) What good are clues if most of them turn out to be wrong? Was there a piece missing in the jigsaw puzzle or not? There was no way to know for sure. The data were there, of marginal statistical significance. Geology is not an experimental science. We could not go out and generate additional reversals. All we could do was look at what we had. There was no way to improve the statistics.

Learning how to pick the right projects is a little bit like learning to pick the right parents. It helps to be in the right place at the right time. But as I thought back on my career, I realized that my most important contributions had come from recognizing good problems to work on. Luis Alvarez had also made his career by picking the right projects.

The periodic geomagnetic reversals of Dave Raup had an intriguing smell. I finally decided that it would be worth a little of my time to try to think about Raup's new work, to try to see if there was some way of explaining it. What I should do is think hard about the problem, maybe for ten minutes, every few weeks. That way my subconscious would keep working. It was an efficient way to use my mind.

I talked to Luie about Raup's new paper. He had summarily dismissed it. "Just ridiculous," he said.

"But you dismissed their earlier paper in the same way," I pointed out.

"You showed me that I shouldn't have done that, Rich. But this new work is different. It's just nonsense."

Months passed. Even with funding, our Nemesis search program was stalled. Our imaging device, the charged-coupled-device, or CCD, wasn't adequate for the program we had planned. It was our own fault, since the CCD we had was really just a "loan" from RCA, something to use temporarily while their research-and-development program produced better and better CCDs. They would be happy to send us a state-of-the-art CCD any time we wanted, but we had not yet requested delivery. If we waited a little longer, we could get a better one. So most of our initial search for Nemesis had used the loaned CCD. We took a lot of data with it before we realized that its low efficiency severely limited the accuracy of our measurements. We finally asked RCA to deliver a good CCD, and we began the modifications in our equipment necessary to install it.

I had many other distractions. I had recently joined the Committee for International Security and Arms Control of the National Academy of Sciences. We met with members of the Soviet Academy of Sciences twice yearly, in Washington and then in Moscow, to discuss technical aspects of arms control. Suddenly catapulted from being a casual observer in these matters to being someone who was involved in serious discussions with "the other side," I had a lot of reading and studying to do to get up to full speed.

My graduate student James Welch had sudden spectacular success with our small cyclotron, the "cyclotrino," designed to replace the large atom smashers that were necessary for radioisotope dating. Jim was about to graduate and look for a job. He thought it would be difficult, since he needed a place where both he and his new wife, a physicist specializing in atomic physics, could work. They jointly applied to six major institutions around the country, and they received six good job offers each. With Jim gone, Kirk Bertsche would be taking over the major responsibility on the cyclotrino. So I became more deeply involved in this work than I had been in years.

These distractions gave me plenty of excuses to forget Raup's new paper. But my conscience kept nagging, and I spent a little time thinking about it, although not the ten minutes per week I had promised myself. Fortunately, a few years earlier I had heard geophysicist Edward Bullard give a fascinating presentation on the Earth's magnetic field, and I had thought about the physics of it. I had even read a *Scientific American* article. I pulled what I could out of my long-term memory.

The Earth is a weak but big magnet. The first practical use of this fact had been for navigation. Small magnets, suspended so that they could rotate freely, aligned themselves with the Earth's field. Such devices were called magnetic compasses. The end of the magnet that pointed north was called, appropriately, the north pole of the magnet. The end that pointed south was called the south pole. The north poles of two magnets will repel each other, as will the south poles. The north poles will attract south poles, and vice versa. This all makes great sense, but the semantics get confusing when you think about the implications. Because the north pole of a magnet points toward the north, that means that the northern part of the Earth is actually a south pole, from a magnetic point of view. So the North Pole of the Earth is a south pole, and the South Pole of the Earth is a north pole. I loved to confuse my friends and students with this paradoxical statement.

What is the origin of the Earth's field? Scientists had once assumed that the Earth is just a large permanent magnet. But we now know from seismic studies that the core of the Earth is molten iron, perhaps with small amounts of other metals, such as nickel, mixed in. Molten iron cannot hold a permanent field, so people now assume that the Earth is a large electromagnet, a magnet driven by electric current. What is the origin of the electric current? The answer is simple: the Earth's magnetic field. If that sounds circular, well, it is.

Commercial dynamos had been invented in the nineteenth century, and they had solved the problem of efficient generation of electricity by making use of a similar circular principle. Before the dynamo, electric generators moved coils of wire past permanent magnets, causing electric current to flow in the wires. The invention of the dynamo improved on this in a clever way. Some of the current from the generator was brought back to the magnets, and run through electromagnets to increase the magnetic fields. So magnets create current, which increases the strength of the magnets, which increases the strength of the current, and so on. The whole process ran away, growing exponentially, giving huge currents, limited only by the energy supplied to the dynamo. The invention of the dynamo made the subsequent inventions of the electric motor and light bulb practical.

We believe that the Earth's magnetic field arises from a similar effect. A small preexisting field, perhaps due to magnetized rock near the surface, causes currents to flow in the metallic iron core of the rotating Earth. The liquid iron in the core is believed to be in turbulent motion, and this motion causes the currents to twist around, in such a way that they generate more magnetic field. Bigger fields mean bigger currents mean bigger fields, and on and on. The magnetic field of the Earth is lifted up by its own bootstraps, as energy from the moving liquid in the core is converted to field. Many people had worked on this theory, and many details remained to be worked out, but there seemed to be a consensus that the dynamo mechanism was certainly responsible for the Earth's magnetism.

The flips of the Earth's magnetic field had been a bigger mystery to solve. The direction of the field throughout geologic history had been recorded by sedimentary rock, which trapped some of the field in it as it formed. When this was discovered, scientists were amazed to see that the direction of the Earth's field flipped every now and then. Our North Pole, which is presently a south magnetic pole, had once been a north magnetic pole. In a short time, perhaps just 10,000 years, the direction of the Earth's field had simply flipped. The last reversal had taken place about 700,000 years ago. There have been about 120 such reversals in the last 65 million years, since the dinosaurs disappeared.

These reversals were well established. In fact, Walter Alvarez and geophysicist William Lowrie had made a specialty of using them as a tool for geology. The sequence of flips matched with the similar pattern in a distant rock allowed them to determine when the rock was formed. But the origin of the flips was more difficult to understand. The favorite theory argued that the field was inherently unstable, and that the flips were spontaneous. Complex mathematics was used to justify this model, mathematics that turned out to be very similar to something later called the theory of "chaos."

Raup upset all this when he claimed that the flips were associated with the mass extinctions. We had known, from the work Walt had done with Lowrie, that there was no magnetic-field flip at the time of the Cretaceous catastrophe. So the association couldn't be perfect. In fact, all Raup claimed was that the *rate* of magnetic-field reversals showed a 30-million-year periodicity, similar to the one he and Sepkoski had found for the extinctions. He made no claim about individual flips being correlated with individual extinctions.

Since I believed that the periodic mass extinctions were caused by periodic comet showers, I decided to try to see how a comet shower could cause the magnetic field to flip—cause it to flip *sometimes*, that is. I tried a few times in the next several months to come up with some physical model, but I made no progress. Two months passed without my giving the subject serious thought.

Raup visited Berkeley, and I told him about progress in our Nemesis search and about calculations I had done on the duration of comet showers. It all seemed to be fitting together, I said, except, of course, for his magnetic-reversal effect. He used the opportunity to give me a new draft of his paper, in which he had clarified some of his arguments. He had also plotted the magnetic-reversal rate. I stared at this new plot. There they were, peaks sticking up every 30 million years. It didn't take mathematical analysis to see them. They were right in front of me, daring me to ignore them, knowing that I couldn't.

What could reverse the Earth's magnetic field? The next day I again decided to spend a few minutes thinking very hard. If another extraterrestrial magnet came by, and its field was strong enough, it could force the Earth's field to flip. The Earth's spin would not have to reverse, just its field and electric currents. But what could bring a strong external field near the Earth? Could comets have such a strong field? I considered this idea briefly, and then rejected it. If comets had a strong field, it would have been detected in the spectra of the gas coming from the comets.

The only thing in the solar system that has strong fields is the sun. Was there some way the strong field of the sun could be transported to the Earth? What happened to the sun during a comet shower? If a few comets hit the Earth, how many hit the sun? I quickly calculated the answer: a few million, one per year during the million-year duration of the comet shower. The big number surprised me. Why hadn't I calculated it before? What effect did a million comets have on the sun? Their mass is small. But the mass of the sun's atmosphere is small, too.

Sunspots have exceptionally strong magnetic fields. With a million comets hitting the sun, thousands would hit sunspots. The comets would vaporize as they got close to the sun, and the plasma could trap the strong field. Maybe the vaporized comets would be blown back toward the Earth, carrying the strong magnetic fields with them. Then I had an important insight. It wasn't necessary for the external field to *reverse* the Earth's field. It would be sufficient if it just *disrupted* the field. The dynamo would turn the field back on, but it had a 50% chance of turning it back on in the opposite direction. Thus it had a 50% chance of causing a reversal, good enough to account for Raup's correlation. And it was just as important to realize that it had a 50% chance of *not* reversing the Earth's field. That could explain why no reversal had been seen at the Cretaceous catastrophe. If the field had been turned off and happened to come back on in the same direction as before, the geologists would not see it as a reversal.

In trying to understand the behavior of plasma near the sun, I felt that I was quickly getting in over my head in theory I knew little about. I called physicist William Press, a friend at Harvard who had made some important contributions to solar theory, and described my idea to him. He promised to think about it. An hour later, in discussing the idea with astronomer George Field, another theoretician at Harvard, they had found the fatal flaw. He didn't call me back until the next day, and by then I had found it myself. The comet would not stay in the vicinity of the Earth long enough to affect its field. Magnetic fields take thousands of years to penetrate the iron core of the Earth, and there was no conceivable mechanism to keep the comet plasma nearby for that long.

The idea hadn't worked, but I thought it was sufficiently clever that I might be able to salvage it someday. I included a brief outline of it in a review paper I was writing for an astronomy conference. But when the referee read the idea, he dismissed it as "totally ridiculous." I should delete it before publication, he insisted. I decided that he was right.

Another month passed. I was flying to San Diego to attend a meeting at which I would see Press, so I decided that it was time to think about the problem again. What about the crater from the impact? Could its formation somehow disrupt the dynamo? I calculated that an impact released enough energy into the Earth to overwhelm the energy in the Earth's field. But how would that energy do the right thing, turn off the field? Could the flow from the shock wave disrupt the dynamo? I knew a little about shock waves, but not enough. It took me about twenty minutes to convince myself that the shock wave could do the job.

Once again I tried my idea out on Bill. It took him about twenty minutes to convince me that it wouldn't work. Even with the fluid flow in the core of the Earth totally disrupted, the inductance of the field would make the electric currents and magnetic fields persist until the fluid flow returned to normal.

But now Press had started thinking. What about the effect of the new crater on the spin of the Earth? It would displace the center of mass, and make the Earth wobble. It would wobble by only 1 centimeter, not enough to do anything. Press suggested that maybe the wobble would grow. This time *I* was skeptical. How could the spin of the Earth be so unstable that wobbles would grow only when craters were suddenly scooped out, and not at other times? He had no good answer, and we eventually gave up.

Two more months passed. Once again I decided to think about the problem. I had no new ideas. Maybe I should solicit help *before* I went any further. Maybe someone else could contribute a less crazy idea than the ones I had been coming up with. I had a meeting with David Shirley, the director of the laboratory, to talk about long-term funding for our work. At the end of our meeting, I said to him, "Your son is a geologist, isn't he? Let me tell you about an interesting paper." I told him about the Raup result. The next day I delivered to him a copy of Raup's paper, at his request, to show his son.

On Wednesday, the ides of August 1985, we were planning to have Saul Perlmutter report on the progress of the Nemesis search at our weekly group meeting. Unfortunately, the computer had been down all week, and Saul had nothing new to report. So at the last minute I decided that this might be a good opportunity to inform my entire physics group of my need for help. I barely had time to make a transparency of Raup's magnetic-field-reversal plot, the one with the "obvious" 30-million-year peaks, to show at the presentation.

Luie came to my talk, and his main contribution was skepticism. He said there was no statistically significant effect. He didn't see any peaks with 30-million-year spacing. Those things that I called "peaks" were simply statistical fluctuations, and he thought I was wasting my time. I responded by reminding Luie that I knew I couldn't convince anybody else that the data were worth paying attention to, but I had managed to convince myself enough so that it was worth a few hours a week to work on the problem. I went on to describe the failed theories I had come up with. I felt like I was reenacting the kind of presentations I had made to Marc Davis and to Piet Hut about periodic mass extinctions: Here are my attempts. Can you add anything new? Can you salvage them?

I described everything I could think of, including the small effect of the displacement of the center of mass of the Earth by 1 centimeter (equal to the thickness of the boundary clay layer), and the potential wobble that the offset would cause. I mentioned that the change in the moment of inertia of the Earth would speed the Earth up slightly, but the effect would be negligible: only a minute per year. Later I discovered that the true speedup would be even smaller, because I had wrongly assumed too large a crater volume. A more accurate answer is a few seconds per year. These days, when we have "leap seconds" on New Year's Eve, we could all notice the effect. But it was far too tiny to affect the dynamo deep in the Earth.

After I gave my talk, one of the physicists in my group, Donald Morris, came over to me to make some comments. He suggested that although a minute per year sounds like a small amount, after a few hundred thousand years the core of the Earth would have lagged behind by many degrees. Wouldn't this twist up the magnetic field, and perhaps force it to turn off? I thought it was a wonderful idea, something new, something I hadn't thought of. But after a few minutes of analysis, we decided the idea had serious problems. The time it took the field to twist up was too long, and the magnetic field would diffuse through the core of the Earth in this time, undoing the twist. I didn't care too much that the idea didn't work. Don and I had spent ten minutes throwing ideas at each other, covering a tremendous amount of ground very rapidly. He had found a gap in my reasoning. I shouldn't dismiss the slight speedup of the Earth's spin as negligible. The progress was exciting. I didn't want it to stop.

Suddenly I remembered an idea I had once had, which hadn't been useful for anything before. There could be a much bigger effect, I told Don, if the nuclear winter effect redistributed a large amount of ocean water to the polar regions as ice. "That would really change the moment of inertia of the Earth," I said. "And get it spinning really fast."

"Oh, that's brilliant! Really brilliant!" Don responded, with what I thought was quite a bit of exaggeration. Now we were both excited. How fast would be "really fast"? If the crust of the Earth were accelerated to speeds comparable to those found in the liquid core (not including the Earth's rotation) that would be fast enough. I seemed to recall that those velocities were about 1 centimeter per second. (Once again, I was wrong.

The true velocity is probably more like a half-millimeter per second, twenty times slower than I had guessed.) How much would the ocean level have to drop to speed up the rotation of the Earth by this much? We quickly calculated that it would take a drop of 100 meters. I seemed to recall that drops this large had occurred. This time my memory was right.

It looked as if we had found a mechanism that might work, and it didn't involve anything crazy. The physics was really quite simple. It was the old spinning ice-skater effect. An ice skater, to spin faster, pulls his arms closer to his body. The Earth, when it pulls its water in closer to its axis (that is, from its equator to its poles), will similarly speed up its spin. Initially only the outer part of the Earth, the solid parts known as the crust and mantle, would speed up. The fluid core would lag behind, like a well-greased axle. The difference in velocity between the inner and the outer core would generate a shear velocity in the liquid iron (the grease of the axle). The new velocities would disrupt whatever was going on before, including the dynamo. When the dynamo turned itself back on, it had a 50% chance of showing a reversal.

It looked like the first "natural" mechanism to link impacts with magnetic-field reversals, but we had to check our numbers. What was the density of the mantle? How fast were the fluid currents in the core? That afternoon and evening I searched books and encyclopedias for better numbers. One critical book had been checked out; when I asked the librarian to trace it, I was embarrassed to learn that it had been checked out to me! I had lost it. It served me right. I went to tell Don how ironic it was that I had lost the very book I now needed. He smiled, picked a book off a shelf, and handed it to me. It was the missing book. He said I had loaned it to him a year ago, and he had never gotten around to returning it.

The density of the mantle was slightly higher than I had guessed, and that weakened the effect. But the next morning I found some numbers for the fluid-core flow velocities, and they were a factor of 20 less than I had guessed. That made the effect so large that the magnetic field of the Earth should be turned off even with ocean-level changes as small as 10 meters. And I had found an article in *Science* showing that 10-meter ocean changes have occurred every 2 million years or so; not quite as frequent as the magnetic reversals, but very frequent indeed. And, what was most important, the drop in sea levels had been extremely abrupt, for reasons not fully understood. Of course, I was sure I knew why they had been so abrupt. They had been brought on by the nuclear-winter effect, following the impact of a comet.

I told Luie about the new theory Don and I had created. He listened thoughtfully, and then said, "That's great, Rich, one of the best things you have ever done!" It was another one of his characteristic turnabouts. He loved the theory. It was beautiful, elegant, nothing forced. Later that day he gave similar congratulations to Don. Friday morning, two days after we had created the theory, we started writing the paper. We had to get things written so we could search for flaws. We would show the paper to real experts, physicists who had worked on the theory of the Earth's dynamo. But I was beginning to wonder whether another *theoretical* paper would ruin my reputation as an experimentalist.

We later refined the theory. We realized that the shear across the liquid core would transport cold fluid above hotter fluid. The cold fluid would try to fall and the hot fluid rise, and they would run into each other. The pattern of flow in the core would be dramatically changed, and it would no longer support the dynamo generation of the Earth's field. Eventually the fluid flow pattern would settle down, and a new dynamo would build up, but the Earth's magnetic field would then have a 50% chance of being in the opposite direction.

We had many problems left to address. Why had there been no magnetic reversals during the last 700,000 years, a time when huge glaciers were advancing and retreating across the northern continents? That answer was easy to find. These recent ice ages had been too slow in their onset. Unless the sea-level drop of at least 10 meters took place in less than a few hundred years, the shear across the liquid core wouldn't be enough to affect the dynamo. How fast were the quicker sea-level drops? We called geophysicist Peter Vail, who had discovered the drops while studying seismic records at Exxon. All he could say was that they were faster than 10 meters in 100,000 years. He suspected they had been *much* faster, but he had no data to prove or disprove that.

Don and I finished a draft of our paper and mailed a copy to *Science* for publication. Simultaneously we mailed copies to all the experts on geomagnetism we could think of, soliciting their criticisms and advice.

As soon as I told Walt about the new theory, he told me of a related, previously unsolved mystery: There were microtektite layers at three of the last four reversals, discovered by Billy Glass. We read Glass's papers and called him. The layers of microtektites had been found in dozens of seafloor cores, and just as Walt had said, they were precisely at the geomagnetic reversals. Glass had published the correlation between the microtektite fields and the geomagnetic reversals as early as 1967. He had told many people about the correlation, and everybody had found it exciting, at first. But then they always asked him to explain how impacts could have caused reversals. Glass had no answer, and said that most people immediately lost interest when they found he couldn't supply an explanation. Another statistical fluctuation, they probably concluded.

We could now supply the explanation. Microtektites were the splash of a comet impact, and we knew how impacts could trigger geomagnetic reversals. It was almost as if a prediction of our theory had been verified.

Don asked Glass about the one microtektite field that didn't have a reversal. Our theory predicted that there should be a geomagnetic excursion at that point, that is, the Earth's field should turn off, as if it were reversing, and then abort and return to its previous value. Every large impact should have either a reversal or an excursion. Glass replied that he thought that there *was* an excursion at the time of the fourth microtektite field, and that he had requested funding to make additional measurements. But his request had been turned down. His work was too far out of the mainstream of scientific research. There were other proposals to be funded, ones that *everyone* felt were important.

Don came to me one afternoon with a question that was bothering him. How could a small impact trigger a nuclear winter? The Cretaceous catastrophe was triggered by an impact that made a crater 100 miles in diameter, and the material that had been lofted was enough to form a centimeter-thick layer of clay around the world. The crater associated with one of the microtektite fields was only about 6 miles across. The clay from a crater $^1/_{20}$ as big as the Cretaceous crater would make a layer only $(^1/_{20})^3$ as thick, only a micron, $^1/_{25}$ the thickness of a human hair. There wasn't enough dirt thrown up to block sunlight. I responded that perhaps the fireball could set extensive fire storms, and the soot could do it. Don thought that was a clever idea. I didn't realize at the time that he was not familiar with the Anders discovery of soot at the Cretaceous boundary, which had not yet been published.

Luie had shown me the preprint by University of Chicago geologist Edward Anders and his colleagues Wendy Wolbach and Roy Lewis reporting the unexpected discovery of soot in the clay. It had been missed by Frank and Helen because their neutron activation method was insensitive to carbon; besides, a nuclear technique couldn't have distinguished carbon in soot from carbon in its more common form of limestone, which is made of calcium carbonate. Anders found that the layer was between 0.3% and 0.5% soot. To make this much soot, an amount of carbon equal to that in 10% of the entire biomass covering the Earth today must have been burned. The fires could have been set by radiant heat from the fireball, or from the hot rock and ashes thrown worldwide by the impact. It was an exciting, fantastic discovery. Anders and his colleagues wrote that their find "confirmed" the Alvarez impact theory, although it added another killing mechanism to a catastrophe that already had more than it needed. The true importance of the Anders discovery lay in the fact that it showed that an experiment with extensive burning of the Earth's surface had already been performed, by Nature, and it made the reality of the nuclear winter more compelling. On October 11, 1985, their paper was finally published in *Science.*

Frank was delighted with the Anders discovery. "Anders really deserved a big one," he said, meaning a big, important discovery in this field. Anders had been a referee on the original iridium discovery paper, and had signed his name to the referee report rather than remain anonymous. Frank continued, "He knew so much more about meteoritics than we did that he really could have chopped us up. Instead, he gave us three pages of corrections, and recommended that the paper be published."

The work of Anders and his colleagues had shown how a relatively small impact could have a major effect on climate, so Don and I thought there were no serious problems left with our paper. Then Raup sent Walt an advance copy of a "News and Views" column to be published in *Nature*, in which he withdrew his claim of periodicity in the magnetic reversals. How could he do that? I spoke to Raup on the phone. "It was a statistically weak effect," he told me. Yes, I already knew that. But that didn't mean it wasn't real, I argued. I told Raup about the new theory Don and I had found. It could explain his effect. Impacts *should* cause reversals. He thought it was interesting, but he decided to go ahead and publish his retraction anyway. He was acting in the best tradition of science, openly withdrawing a result that he no longer believed in. Maybe his level of skepticism had just shifted upward slightly. He still stood firm on the periodicity of the extinctions. Maybe he was withdrawing his weaker effect so that he wouldn't be branded a "nut," and so his more important discovery would receive proper attention. But it is difficult to know what is most important when you are in the middle of things.

16. Searching

OUR TWO papers were published in the April 19, 1984, issue of *Nature*. The first one was "Extinction of Species by Periodic Comet Showers" by Marc Davis, Piet Hut, and Richard A. Muller. The editors had (without our permission) taken the footnote suggesting names for the companion star and promoted it to a paragraph in the paper, deleting all the names except Nemesis. The second paper, by Walter Alvarez and me, was "Evidence from Crater Ages for Periodic Impacts on the Earth."

At a subsequent conference in Boston, I gave a talk about our search for Nemesis. I predicted that it would take only three more months to look at all the dim red stars in the Northern Hemisphere. After my talk, Bill Press and Harvard physicist Paul Horowitz told me that such bold predictions were unnecessary.

"Say a year," they advised. "You can't do anything in three months." I retorted, "But I think that three months is how long it will take."

People at the conference asked me if we knew what direction in the sky to look. I answered that we had looked at the effects that Nemesis might have on comet orbits to see if they might indicate the position, but the effect of Nemesis was negligible compared with that of passing stars such as Alpha Centauri, so we could tell nothing about its direction. I said that the most likely constellation for it to be in was Hydra. Why? Simply because Hydra had more square degrees than any other constellation, so it had the greatest chance. I got a mild laugh at my facetious answer. Piet had looked at the stability of the Nemesis orbit and concluded that it was slightly more likely that Nemesis was near the plane of the Milky Way galaxy. For a fixed-period companion (26 to 30 million years), an orbit in the plane had the smallest diameter, and was therefore most immune to the perturbations of passing stars. (I had earned an acknowledgment on that paper by pointing out to Piet that the gravitational force of the galaxy tended to pull the sun and Nemesis closer together, rather than rip them apart.) But Piet's analysis did not help us in the search, since most candidate stars were also in or near the galactic plane.

The editors of the *New York Times* decided that Nemesis didn't really exist. They said so several times in their editorial pages, continuing in the tradition they had begun with the anti-Goddard, antirocketry editorial of 1921. The most bothersome of their editorials was entitled "Miscasting the Dinosaur's Horoscope." After summarizing the impact discovery, the periodic extinctions, and the Nemesis hypothesis, the editorial began its criticism by stating some old discredited claims as if they were new discoveries:

> On closer scrutiny, the alleged repeating pattern of mass extinctions has faded. The dinosaurs and other vanished species didn't all turn feet up in a day; some were in decline before the end of the Cretaceous. The thin layer of iridium that has been found in many geological strata dating from 65 million years ago could indeed have come from a meteorite, as the Alvarezes suggest, but eruptions of volcanos are now known to be sources of iridium too.
>
> Terrestrial events, like volcanic activity or changes in climate or sea level, are the most immediate possible causes of mass extinctions.

They concluded with this remarkable sentence:

> Astronomers should leave to astrologers the task of seeking the cause of earthly events in the stars.

"Like the tides," I said to myself. Secretly, I was delighted that the *New York Times* had taken an anti-Nemesis stand. The worst thing that can happen to a new theory is to have it ignored. Had the *Times* taken a pro-Nemesis stand, my colleagues would have complained that I was publishing my results in the popular press. By taking an anti-Nemesis position, the *Times* was guaranteeing that the theory would get attention and I would get the *sympathy* of my colleagues.

Walt pointed out that we could evaluate the opinions of fellow scientists by the way they referred to the star. Skeptics referred to it as "the death star," or even as "the putative death star." Those who were somewhat undecided and neutral referred to it as "the proposed solar companion star." The true believers called it "Nemesis."

I came back to my office one afternoon and was greeted excitedly by Peter Friedman, one of my graduate students. He had taken a message for me from Dave Cudaback. The scientific grapevine had been at work. Dave had heard from Chris McKee about a rumor spread by Martin Cohen at the NASA Ames Research Laboratory. Cohen had heard that two astronomers had found Nemesis.

The astronomers were Frank Low, from the University of Arizona, and Thomas Chester, who worked for the Jet Propulsion Laboratory at the California Institute of Technology. Together they had been studying the data obtained by IRAS, and they had found a bright object in the data that matched a 15th-magnitude star photographed at Mount Palomar using the 200-inch Hale telescope. It had no detectable proper motion and no visible spectral lines. These were characteristics that Nemesis would have if it were a light-mass, "brown dwarf" star, too light for thermonuclear fusion to have ignited in its core. The team was refusing to give the position of the star until they completed further measurements.

I immediately called Chester. He wasn't in. I left a message asking him to call me back.

Saul Perlmutter came by, and I told him about the reported discovery. He had been leading the effort by our group to find Nemesis. Saul, Peter, and I tried to decide whether we should be happy or disappointed. The identification of Nemesis would be a great discovery, and an important and simplifying focal point for all future work on the catastrophes. But it hadn't been found by us. We could console ourselves with the thought that if the star really had the properties rumored, a temperature of 1200 degrees Kelvin and a visible magnitude of 15, then it was too dim for us to ever have found, even if our system was fully operational. We could find red dwarfs, but not brown ones. We all decided that if the rumor turned out to be true, and Nemesis had been found, that we would be happy. We would even celebrate.

Chester finally returned my call. I began, "Hi, Tom. I just heard a rumor that someone at Cal Tech has found a candidate for Nemesis. Do you know anything about it?" I could hear Tom chuckling at the other end of the line. "Yes, we are the ones. But don't get excited. Frank Low and I have just completed a week of careful measurements on the star. Just yesterday we were able to show it is *not* Nemesis. It turned out to be more than 6.5 light-years away. We found spectral lines in the infrared. It is a carbon star." A carbon star is a large red star, made to appear dim only by virtue of its distance.

I felt relieved, and then I realized I wasn't supposed to feel good about the discovery's turning out to be wrong. Since I was going to celebrate the discovery, I should have felt sad. But I didn't.

One night, as I was lying in bed, I realized that an open star cluster could trigger periodic comet showers if the star cluster just happened to be situated in the right place in the sky. It would have to be at one of the "galactic poles," that is, directly above or below us in the plane of the Milky Way galaxy. It would oscillate up and down in the galactic plane, just like the sun, and the solar system would pass through it twice every oscillation. One of the stars in the cluster would be sure to trigger a comet shower. I got out of bed and looked for a book of star maps. In *Norton's Star Atlas* I found the galactic north pole, and right smack on top of it was one of the largest nearby open star clusters, Coma Berenices. There was no need for Nemesis now! Collision with Coma Berenices was a simpler, more elegant theory than the Nemesis hypothesis, because this cluster was already there. It didn't have to be found. I felt very stupid. Why hadn't I thought of this before? Why had nobody else thought of this before? What should I do next? Go back to bed.

In bed I found that my brain had lost all interest in sleeping. It just kept thinking. I had always said that I would abandon the Nemesis hypothesis as soon as anybody found a better theory, and I had just found it. There must be some way to prove that this new theory was right. I remembered the exciting moments when Frank Crawford and I had concluded that the Alpha Centauri system was orbiting the sun. Proper motion had ruled out that theory. What about proper motion of the cluster? I realized it had probably been measured. Proper motion of star clusters had been used as an important indicator of galactic distances. I popped out of bed and went back to my astronomy books. On page 279 of Allen's *Astrophysical Quantities* I found entries for the velocity of the cluster. It is moving with a velocity of 8 kilometers a second in a direction that will move it far from the path of the sun by the time we get there. The sun and its comet cloud will never get near the stars in the cluster. The new idea was wrong.

Proper motion ruled it out. I had fallen in the same trap twice. Once again the Nemesis hypothesis was the only surviving theory.

My doubts rose again when I received a paper from Dan Whitmire, the astronomer who, along with Al Jackson, had proposed a companion star to the sun at the same time Davis, Hut, and I had. This time he had a new coauthor, John Matese, and a new theory, which he called "Planet X." It was an outgrowth of the tentative theory that he had told me about at the Berkeley conference.

The Planet X hypothesis was even cleverer than the Nemesis theory. A tenth planet, as yet unobserved, had previously been proposed by others to explain slight deviations observed in the orbits of the outer planets. What Whitmire and Matese had realized was that the same tenth planet, Planet X, could perturb the *inner* part of the Oort comet cloud, and cause periodic comet showers. It turned out not to be difficult to set the orbit size and shape in such a way that the perturbation would take place every 26 to 30 million years.

They assumed that Planet X had a mildly eccentric orbit that was not in the plane of the solar system, and that the inner part of the comet cloud was flattened into the solar-system plane. This assumption was ad hoc, but it was reasonable. The outer comets had been stirred up by passing clouds, but not these inner comets.

The orbit of Planet X normally kept it well away from the comets, until the mild perturbations of Jupiter and Saturn caused it to scrape the edges of the comet cloud at perihelion and aphelion. The comets would be scattered by Planet X; some of them would fall into the orbits of the outer planets, and they would be scattered further. Eventually some of them would reach the region of the Earth.

It was an elegant theory, maybe even better than Nemesis. I decided I had better look at this theory carefully. Maybe it was even true, and we should be searching for Planet X rather than for Nemesis. I hoped not. I thought "Nemesis" was a better name than "Planet X."

I took the paper home and read it carefully that evening. I tried to reconstruct their model and do the Whitmire and Matese derivations on my own. I kept running into one problem. As the comets were scattered, first into the outer solar system and then into the inner one, and finally to the orbit of the Earth, the storm seemed to spread out over a long period. It looked so long that there would be no special time when the probability of being hit by a comet was appreciably higher than at any other time. I could find no place in their paper where they addressed this issue. I finally decided that it might be an error in their paper. I called Whitmire the next day and told him about my calculations. He was very receptive to the criticism and said he had no immediate answer. He promised to get back to me. I was pleased that my comments weren't obviously stupid.

A few days later he called back and said that my criticism was indeed valid, and their *old* theory was in fact wrong. But they now had a variation on the theory that didn't have the same weakness. They had a revised version of the paper, and had added an acknowledgment to me, as well as to astronomer Paul Weissman, who had independently pointed out the error to them.

I asked myself why I hadn't attempted to salvage their old theory, rather than just knock it down? I realized once again that I had been getting lazy. I had a theory of my own, and I was trying to disprove other theories. I wasn't trying to find alternatives that worked. True, I had thought of the Coma Berenices model, but I had made no effort at all to try to salvage the Planet X model. I looked forward to receiving the revised paper of Whitmire and Matese.

I first saw their new paper at a conference in Arizona, which most of the major participants in the periodic-extinction business attended. Don Morris and I were sitting next to each other when Whitmire presented the revised Planet X theory. Don saw a problem with the new version. Whitmire and Matese had to assume that the inner part of the comet cloud was flat, like a disk, and that Planet X moved up and down out of the plane of this disk. Planet X perturbed the comets only when its orbit precessed in such a way that at the extremes of its motion it scraped the edge of the comet disk. But Don realized that Planet X would continue to perturb the comets even when it was far away, and would cause them to spread out of the plane. If that were the case, then the comet storms would not come at well-defined times, but would be spread out over tens of millions of years, in contradiction to the mass-extinction data.

I spent about ten minutes trying to calculate the effect. I finally succeeded in deriving an equation that showed that Don's criticism was correct. I whispered to Piet, who was sitting on the other side of me, that Planet X would perturb the comets out of the plane. I wanted to show him my calculation, but he said he would prefer to think about it. About one minute later he turned to me and said, "Yes, you are right." He then gave me the formula, out of his head, describing the time for the comets to be disturbed. It was the same as the one I had taken ten minutes to calculate with pencil and paper.

Don and I later showed the result to Whitmire and Matese, and they didn't have any immediate answer to the new criticism. We then showed it to astronomer Scott Tremaine, who was supposed to deliver a summary paper at the end of the session. Scott agreed that the problem looked serious, but in his summary talk he simply said that the Planet X theory, the only viable alternative to the Nemesis theory, was too new for us to be sure it didn't have serious flaws. He also said that he thought it more likely that a statistical fluctuation would eventually account for the "periodic" mass extinctions, rather than a putative death star.

Don telephoned Whitmire a few weeks later, and Whitmire admitted to him that the perturbations pointed out by Don made the comet storms dwell for a long time, but he wasn't yet convinced that the extinction data truly required short storms. A few weeks after that I spoke to Piet again. He and Tremaine had both concluded that the criticism was sufficiently severe that the Planet X theory "was as dead as a theory could be." I interpreted that to mean that Piet wouldn't work on the theory any more, but would leave the theory to Whitmire and Matese to revive. In a sense it was a shame. The Planet X theory was elegant, perhaps more elegant than the Nemesis theory. Piet quoted Thomas Huxley: "The great tragedy of Science: the slaying of a beautiful hypothesis by an ugly fact."

There was other work to be done besides looking for Nemesis and considering alternative theories. There was still a great deal of information hidden in the record of the sedimentary rocks. Luie's new iridium detector was under construction. Initial tests of its components made by Frank indicated that it would operate right up to specifications. Frank and Helen hoped to be able to measure 5,000 samples per year. This system would answer the questions of whether there were iridium peaks at all the extinctions and whether any of the peaks were multiple, indicating a comet storm.

In the meantime, Walt was still studying the geophysical evidence in the rocks. His proposal for funding had been turned down a year earlier, and this year, somewhat piqued, he refused to rewrite it, merely resubmitted it unchanged. There was nothing wrong with it last year, so why should he change it? Finally, he received good news. This year the National Science Foundation would fund him, but at a greatly reduced level.

Why did the paleontologists believe so strongly that extinctions were gradual? Walt decided to look at some of the evidence. A few days later he called me and said he had something very interesting to show me. He came up to the laboratory later that day. With him he brought a copy of an article by Gerta Keller analyzing the foram extinctions near the Eocene boundary 35 to 39 million years ago. Keller was an excellent paleontologist, and had collected a large set of data on the disappearance of microscopic fossils. Walt showed me a plot in which she had drawn a set of lines, one for each species, in the order in which they had disappeared. The end point of each line showed when the creature was last seen. About half the species seemed to go out gradually, but the other half seemed to disappear in three separate, abrupt catastrophes. "Looks like clear evidence of a comet storm to me," I said.

"Yes," Walt said, "that's what you might think at first. But let me show you a subtlety. See those wiggly lines that Keller has plotted right at those times you called 'catastrophes'? Each one indicates the presence of a hiatus."

"Hiatus." That was a word that Walt had taught me before. It meant missing rock. While the sediment is compressing in the seabed to form limestone, there are many possible phenomena that can wash it away, including a tilting of the bed or a storm whose effects reach the bottom. If a few meters of rock were missing, then all the species that had disappeared gradually in the missing rock would seem to have disappeared abruptly. So it wasn't real evidence for comet storms, after all. In all the regions where the rock sample was complete, the species seemed to vanish gradually. Walt told me that this was the kind of evidence paleontologists used to show that extinctions were gradual.

It looked convincing, but I still wondered why Walt had wanted to see me right away. Had he become convinced that the comet-storm hypothesis was disproven by these data?

"I began to wonder how Gerta determined that a hiatus was present at these locations," Walt said. "It usually isn't easy to detect missing rock. You can sometimes do it by finding stratigraphy elsewhere that matches. I wondered how she had done it for the Eocene and Oligocene. Fortunately, I didn't have to wonder very long. Gerta is really a good scientist, so she had described in her paper how she had done it. Do you want to guess?"

Now I was beginning to see Walt's point. "By the abrupt disappearance of many different species?!" I blurted out.

"Exactly!" said Walt.

Keller had taken data that could have been interpreted as evidence for a mass extinction and used it as an indication that there was missing rock. There was no independent evidence that there was rock missing. The

assumption of gradual extinctions had been her *starting point*. It was part
of the paleontologist's paradigm. To be fair to Keller, she did not conclude
in this paper that the extinctions were gradual. She was very clear about
gradualism being one of her assumptions. It was other paleontologists
who referred to her data as evidence of gradualism, and it was circular
reasoning. One scientist makes an assumption in order to facilitate the
analysis of the data. Other scientists, unaware of the assumption, use the
data as evidence of gradualism. "Nobody reads the literature any more,"
Ed McMillan had said. I had first learned of circular reasoning in seventh
grade. It had seemed like a trivial trap that no intelligent person would
ever fall into. It had turned out to be much more subtle when several
different scientists were involved.

Walt was careful to point out that we still didn't know whether the rock
was missing or not. With our new paradigm of catastrophic extinctions,
we could no longer assume that abrupt disappearance of many species
indicated a hiatus in the rock. But likewise we couldn't be sure that there
wasn't a hiatus. Additional measurements would have to be made. A lot of
work was ahead of us.

Several months later, just after he came back from a visit with Erle
Kauffman, Walt called to say he had more exciting news. Kauffman was
the self-proclaimed gradualist, the paleontologist who had convinced me
at our Berkeley conference that the corals had become extinct tens of
thousands of years before the dinosaurs. He had been one of the original
skeptics of catastrophic extinctions, but had been slowly won over. Now
he seemed to have become a leader in a movement toward accepting not
only abrupt mass extinctions, but stepwise extinctions, the kind that our
comet-storm theory predicted. Kauffman now believed that "we are only
one to two years away from a general theory of mass extinctions."

Kauffman had told Walt about new work by Thor Hansen on the
marine extinctions near the Eocene boundary, the same boundary that
Keller had analyzed. Hansen had found three major sudden extinctions
near this boundary. Walt gave me the impressive numbers. Between the
late and middle Eocene (about 42 million years ago) 385 species of
mollusks existed. Suddenly (meaning over a period of less than 50,000
years), 89% of the gastropod (snail) species and 85% of the bivalve
(clam) species disappeared forever. About 2 million years later, in the
middle of the upper Eocene, the mollusks had made a strong comeback.
There were then 273 species. But in a second catastrophe, 72% of the
gastropods and 63% of the bivalves died. And, finally, at the Eocene-

Oligocene boundary, in the space of a few meters, of 86 species, 97% of the gastropods and 89% of the bivalves died. There had been three massive extinctions.

"That's a total span of 4 million years for the comet storm; a little long," I said to Walt.

He replied, "That's only if you accept the standard time scale. Sandro has made his own careful measurements, and he's concluded that the period described was more like 2 million years instead of the 4 million years. That fits right in with the Nemesis model."

The next day I went to my file and retrieved a copy of Keller's paper. I looked at where she had marked each "hiatus." They agreed qualitatively with the times of the mass extinctions found by Hansen. Walt had warned me to be cautious. Hansen's data hadn't been published yet. They might change. And we didn't have accurate enough time scales to correlate Hansen's extinctions exactly with Keller's hiatuses. But I couldn't help feeling optimistic. Everything seemed to be fitting together beautifully.

Frank called me. "Rich, remember that rock you brought back from Italy, at the bottom of the Bonarelli?" I had taken a trip, almost a pilgrimage, the previous summer to visit the site where Walt had first found the iridium level. Walt had showed me another black layer, 90 million years old, the Bonarelli. He had previously sampled several places in this layer, but had not found an iridium enhancement. I suggested that we sample at the bottom of the layer, and Walt and I had brought these samples back for Frank and Helen to measure.

"We just finished checking it. There's an iridium enhancement there."

17. I Think I See It

"WHICH ONE is Nemesis, Daddy?" My seven-year-old daughter was lying in her sleeping bag staring at the stars sparkling in the sky above. We were in the midst of a four-day backpack trip in the Sierra Nevada, just outside Yosemite. "I don't know, Betsy. We haven't found it yet. But we're pretty sure it's not one of the brighter stars." Betsy was undaunted. Suddenly she said, "I think I see it!"

I wish it were that easy. Eighteen months earlier, I had been overly optimistic and had predicted that it would take three months to find the star. "It is a scientist's *duty* to be optimistic," Edwin Land was fond of saying. The search is taking years, not months, but our enthusiasm and optimism haven't waned. We have already ruled out all of the naked-eye stars, Betsy notwithstanding.

Nemesis, if we are right, is currently as lost as a needle in a haystack among a million brighter stars. But when you find a needle, you can tell that it ain't hay. If we knew which one it was, we could see it through binoculars. With a small telescope its distance from the sun and its orbit could easily be measured, once we knew which one it was. Some skeptics say the Nemesis theory is pure speculation. They won't pay attention until we have found the star. I admit that I can't prove for sure that Nemesis is out there, but I think the odds make it a very good bet, good enough to bet several years of my career. First we will finish our star survey of the Northern and Southern Hemispheres. If the mass of Nemesis is greater than about $1/20$ of the sun's mass, we will find it. Soon the Hipparcos satellite will be launched into space with the capability of making a sweeping survey of nearby stars and looking for smaller stars. If Nemesis is not found by this satellite, it will be time to look for another theory. But I am hoping that we will not need the help of Hipparcos.

What *do* we know for sure? We know that an extraterrestrial object, either a comet or an asteroid, hit the Earth 65 million years ago and brought to an end the great Cretaceous period of the dinosaurs. I think this conclusion is as firmly established as any theory can be after half a decade. I can say this freely because my contribution to this early work was negligible. I have watched other great discoveries go from controversy to acceptance, and other great discoveries go from controversy to retraction. This one will not go away. In the words of a court of law, I believe that the case has been proven beyond a reasonable doubt.

I also believe that the Earth is subjected to periodic storms of comets or asteroids. The important discovery of periodic mass extinctions by Dave Raup and Jack Sepkoski lies on firm and careful analysis of the data. I would not claim, however, that this conclusion is as solidly established as the Alvarez impact discovery. It is conceivable that we were just very unlucky, and Nature happened to cause mass extinctions in a way that just *looked* periodic. But I don't think so. The periodic extinctions, and the periodic cratering that goes along with them, are firmly established, to my mind. To prove a case in a civil court it is unnecessary to prove it beyond a reasonable doubt; it is only necessary to show that the preponderance of the evidence supports the case. I believe that this is true for the periodic comet storms.

Nemesis is in a different category. It is a beautiful and simple solution to the mystery of the periodic mass extinctions. I also believe that it is the only theory suggested so far that is consistent with everything we know about physics, astronomy, geology, and paleontology. But the evidence for Nemesis is circumstantial, and it hasn't yet convinced the bulk of the scientific community. It is an elegant theory, a marvelous prediction, that needs verification. We had looked at rock under our feet and predicted a star. We need direct evidence, a smoking gun, a body, the star. When we find it, hundreds of astronomers around the world could verify within a week that the star we have found is part of the solar system, orbiting the sun. (More correctly, the sun and Nemesis would both be orbiting their common center of mass.) It would immediately put an end to most of the controversy that has surrounded the mass-extinction work, because unlike many prior discoveries, it is easily checked. A piece of hay will turn out to be a needle.

The discovery of Nemesis would fill in the last piece of the jigsaw puzzle. Or would it? We've been surprised before. Every time it appeared that the puzzle was complete we were suddenly led in a wonderful new direction. Walter Alvarez had become interested in "an inconspicuous layer of clay in the Apennines." Luis Alvarez had suggested that trace analysis of iridium could be used to measure sedimentation rate, but to everyone's surprise it demonstrated that an extraterrestrial impact had taken place. The clay layer was found worldwide, and analysis showed it was about 10% asteroid or comet material, the rest coming from the vaporized rock thrown up by the impact. Five mass extinctions are now known to have iridium signals. There is the original one at the end of the Cretaceous period, 65 million years ago, and another at the end of the Eocene, 35 to 39 million years ago. More recently, Digby McLaren, a geologist at the University of Ottawa, Carl Orth of the U.S. Geological Survey, and their collaborators at Los Alamos found an iridium layer at the Frasnian-Famennian boundary, 367 million years ago, at the time of a major global extinction in the late Devonian period, and a Polish team has found iridium at the Callovian-Oxfordian boundary, 163 million years old, at locations in Poland and Spain. We found iridium 90 million years old at the Bonarelli; the layer was found independently by Orth. Of course none of the new discoveries has yet been subjected to the scrutiny that has been given to the Cretaceous boundary, so there may be even more surprises awaiting us.

The studies of the effects on climate of dust thrown into the air by the impacts led to the discovery of the nuclear winter. The loop connecting mass extinctions with nuclear winter was closed when Edward Anders found soot in the boundary clay layer, suggesting that vast fire storms had been set by the impact. As Walt once expressed it, "Only gradually have we come to appreciate the appalling violence of the geologic event that produced that thin layer of clay." The attempts of Raup and Sepkoski to show that mass extinctions occur frequently led them to a surprising, and at first totally inexplicable, conclusion: that mass extinctions take place on a nearly regular 26-to-30-million-year schedule. Based on this discovery, Marc Davis, Piet Hut, and I proposed Nemesis, a companion star to the sun that triggers comet storms, a theory simultaneously proposed by David Whitmire and Albert Jackson. Our theory immediately led Walt and me to the discovery that impacts on the Earth follow the same schedule as the mass extinctions, a correlation found independently by Michael Rampino and Richard Stothers. The concept of storms of comets proved to be more general, and testable, than the Nemesis theory itself, and led to the discovery that the mass extinctions were punctuated during the several million years of their duration. A belief in comet storms led Donald Morris and me to a method of explaining some of the geomagnetic reversals.

What else could be out there, having left behind subtle clues, to challenge our observational abilities and intelligence? We know what we are looking for, but in the past such knowledge has not been a very good guide to what we would find. If there was a second star in our solar system, it would have major implications, perhaps some of them easily testable. Which should we spend time thinking about? Could the rings around such planets as Jupiter and Uranus be the debris from comet storms hitting satellites? Did comet impacts have a role in the creation of the Apollo objects, those peculiar asteroids that have Earth-crossing orbits? What are the consequences of all this for the formation of the comet cloud, and of the solar system? What did the early solar system look like? Was the heavy bombardment by comets that took place 3.5 billion years ago due to a closer orbit for Nemesis, and the continuous comet storm that would have accompanied such an orbit? Did such a storm prevent evolution from occurring on the early Earth, or did it trigger it? What secret of evolution is next to be uncovered?

It has not been just an intellectual puzzle, quietly fit together in a drawing room. There have been strong forces all along trying to get us to abandon the puzzle. These include self-doubt, inadequate funding, the feeling that we were making no progress, and the opposition of "experts." A more accurate image is that of an explorer, trying to put together a map of an unknown world, unsure of the value of what he is going to find and how he is going to repay his debts, while suffering from shortages of supplies and attacks by the natives.

Is the puzzle almost complete now, with one tiny piece called "Nemesis" yet to be found? Or have we simply been working a small area of a much larger puzzle, one that will lead to a far grander picture than we are capable of imagining? When Columbus found the West Indies he had no idea that the huge continents of North and South America were looming over the horizon. In the past our imagination has not been as wild as the inventiveness of Nature. What will emerge? There is no way to know.

Index